# THE WORLD'S
# WORST INVENTIONS

# THE WORLD'S
# WORST INVENTIONS
## THE MOST STUPID GADGETS AND MACHINES EVER MADE

**JACK WATKINS**

**amber**
BOOKS

This edition first published in 2010

Published by
Amber Books Ltd
Bradley's Close
74–77 White Lion Street
London N1 9PF
United Kingdom
www.amberbooks.co.uk

ISBN: 978-1-906626-94-5

Project Editor: James Bennett
Picture Research: Terry Forshaw
Design: Zoë Mellors

Printed and bound in Thailand

# CONTENTS

# INTRODUCTION

The world is full of bright ideas and creative geniuses, and of fabulous inventions that make life easier. But for every great invention, there's one that's even more wonderful, not for its usefulness, but for its mere existence. This book celebrates the hitherto unrecognized inventions that are special only for their spectacular ineffectiveness – gadgets, machines or vehicles that were designed with fundamental flaws that made them unfit for purpose, or objects that were created to solve problems you never knew existed.

Under what circumstance, for example, would you seriously think of putting on a pair of two-way shoes, or wearing glow-in-the-dark sunglasses, or fooling around with a square tennis ball? How dissatisfied with your appearance did you need to be to want to buy Mr Trilety's Model 25 Nose Shaper? How impervious to ridicule was the man who wore a moustache protector while eating? Can you imagine being rich enough to squander your money on redundant extravagances such as the Aquatic Pod Suite, or the Energy Cocoon?

**Above:** *Flotation suits made you look more equipped for bath time than a day at the seaside.*

Above: *You are always guaranteed a laugh if you take to the roads on a monowheel.*

Not all these 'bad' inventions merit such sneers, however. It's hard not to feel a sneaking affection for the English eccentric Harold Bate, chugging along country back roads in his car powered from the by-products of chicken and pig manure. You can't help but smile at the valiant, if rather misguided, Herr Richter and his madcap plan to set a new speed record on a 'rocket-powered' motorbike. Richter emerged unscathed, even if the bike did not, but not all such harebrained ventures leave you laughing. How sickeningly horrible, by contrast, was the fate of the Austrian Franz Reichelt, who plunged to his death from the Eiffel Tower in 1912 when his 'parachute overcoat' failed to open?

Some of these gadgets or inventions were simply ahead of their time, while some were obsolete almost as soon as they became available. Others never got beyond the patent stage. But just as we smile at the most farcical of them, we should spare a moment to remember that they were once somebody's dream solution to a problem, however trivial, even if they didn't turn out as planned. It's still possible to admire the human ingenuity inherent in their utter hopelessness. They are certainly a cause of unfailing entertainment.

# CRAZY CLOTHING AND ODD OUTERWEAR

It's no surprise, given the nature of human vanity, that many of the inventions in this section reflect concerns about appearance. But they also provide a snapshot of the times in which they arose. Crinoline dresses, impractical though they now seem, were actually an attempt to liberate the woman of the 19th century from the tyranny of petticoats. The moustache guard is a reminder of the time when cultivating a forest of facial hair was a badge of preening masculine pride. The hair cutter for the bob shines a light back on the days when the style was a craze among young female flappers of the Jazz Age. You can have fun choosing which of these contraptions is the 'worst', but here are two strong candidates – the blueprints for the refrigerating suit and the body-attachable sunshade. While they may have been sketched out with due seriousness, in practical terms, they were as futile as the head umbrella or the two-way shoe.

Left: *Stealth umbrellas are a handy innovation, as long as you know which direction the wind and rain are coming from.*

# HAIR CUTTER FOR THE BOB

When the bobbed hair fashion took off in the 1920s, it was big business for male barbers. Up to this point, female beauticians were more used to dealing with women who had their hair styled or curled, rather than cut. Women turning up asking for a snip were liable to be sent away.

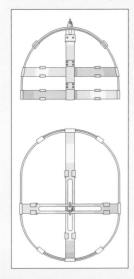

But, of course, cutting a woman's hair in a bob demanded new and exacting skills from barbers, who were more used to thrashing merrily at the scalps of the average, easy-to-please male. D.E. Smith's hair-cutting device must thus be read as a well-intentioned attempt to bolster the nerves of the hesitant barber, suddenly faced with an army of demanding young women customers. Made of iron, full of twiddleable nuts, bolts and screws, you may not have had a clue what you were doing, but you certainly could look professional while you wielded the thing.

**RATINGS**

| | |
|---|---|
| **STYLE COUNT:** | ★★★★★ |
| **ORIGINALITY:** | ★★★★★ |
| **USEFULNESS:** | ★★★★★ |
| **NERD APPEAL:** | ★★★★★ |
| **LONGEVITY:** | Fortunately, there's no evidence to prove that this nutty contraption was allowed to harm a hair on anyone's head. |

*'Literally raises the hair on your head.'*

Left: *Although the arrival of the bobbed hairstyle summoned a new wave of hair-cutting technology, Smith's device looked more like something used by a surgeon about to perform a lobotomy.*

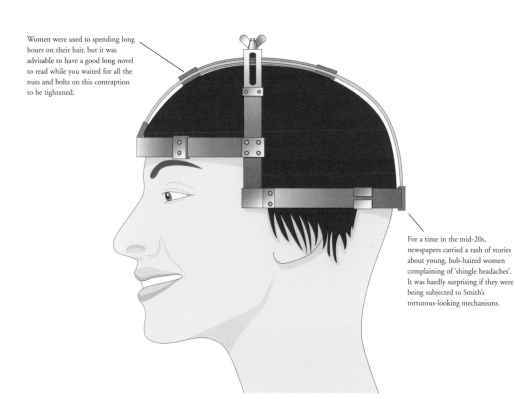

Women were used to spending long hours on their hair, but it was advisable to have a good long novel to read while you waited for all the nuts and bolts on this contraption to be tightened.

For a time in the mid-20s, newspapers carried a rash of stories about young, bob-haired women complaining of 'shingle headaches'. It was hardly surprising if they were being subjected to Smith's torturous-looking mechanisms.

11

# BODY ATTACHABLE SUNSHADE

Charles E. Gill filed his patent for the Body Attachable Sunshade in 1902. His aim, he said, was to provide for cyclists and outdoor workers, or anyone exposed to the sun at the hottest season of the year, with 'a light and easily adjusted protector that can be supported by the body without inconvenience'.

It consisted of a series of adjustable metal tubular shafts, on top of which was an umbrella-shaped canopy. Held in place by screws and an elastic belt strapped around the waist and shoulders, it was extraordinarily complicated to assemble. Neither was it likely to be comfortable or practical especially

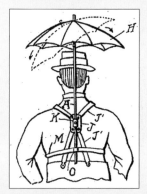

when moving over rough terrain – Gill admitted – because the shade and shaft sometimes waved around or dipped back and forth before being righted by the elasticity of the belt attachment. The diameter of the shade was 60cm (24in) when expanded, and could be adjusted according to the angle of the sun. But anyone hot and bothered by heat would have been driven to spontaneous combustion by the time they had assembled Gill's cumbersome device.

*'It works best if you don't move.'*

Left: 'The circumference of the shade being small, its use will not in the least interfere with the wearer entering streetcars or houses or passing through crowded thoroughfares,' said Gill, optimistically.

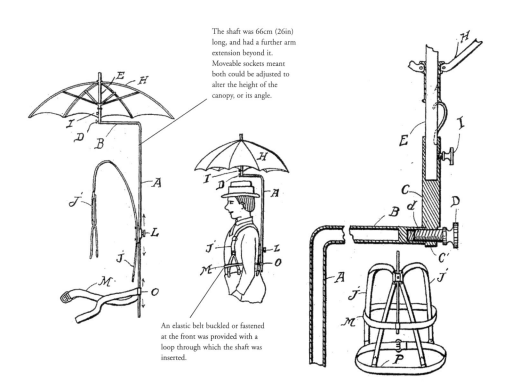

The shaft was 66cm (26in) long, and had a further arm extension beyond it. Moveable sockets meant both could be adjusted to alter the height of the canopy, or its angle.

An elastic belt buckled or fastened at the front was provided with a loop through which the shaft was inserted.

13

# CHANGING TENT

each etiquette can be a tricky matter when it comes to dressing and undressing. In slipping in or out of a bathing costume, people may resort to hiding behind a car door, or just having a partner hold out a towel. It's invariably a clumsy and undignified business. Yet if creating a laughable public spectacle is the chief fear, surely zipping up a personal changing tent in full view of everyone else on the beach is an even surer way of achieving it.

Modern cylindrical body tents rest on the shoulders, and have little openings allowing you to poke your head out the top while you do the necessaries. They are lightweight and collapsible, and certainly less hassle to erect than a full blown changing tent. But it's difficult to imagine bashful types of people – for whom they were intended – wanting to draw attention to themselves in such an obvious way.

## RATINGS

**STYLE COUNT:** ★★☆☆☆ Some of those 50s tents definitely had something

**ORIGINALITY:** ★★☆☆☆

**USEFULNESS:** ★★☆☆☆

**NERD APPEAL:** ★★★★★

**LONGEVITY:** They're funny, but they have remained in demand.

*'Can also be used for rolling down hills!'*

Left: *This early prototype actually was held up by straps attached to a cap. The idea was for the tent to extend all the way down to the ground, but they ran out of material.*

She doesn't look like a
bashful type, does she?

In past times, a lot of people just
made do with an old bedsheet,
cutting an opening for the head.

15

# COCONUT HELMET

There are some inventions you'd love to see succeed but fear that, with the best of intentions, they are doomed to failure. Such a one is the coconut safety helmet. Its inventor is a likeable, sweet smiling academic from Malaysia. He spots a chance to help struggling local villagers by recycling the waste materials from coconuts to make a durable, eco-friendly cycling helmet. It passes every impact test and he travels to an internationally recognized show for new inventions. He's the toast of the event, surrounded by clicking cameras and aimated reporters. But you can only hope he's enjoyed his moment in the spotlight, because who, when all the euphoria fades, is going to want a helmet that smells like a coconut?

*'At least if you get hungry you can eat your helmet.'*

Left: *Most inventions arise from 'thinking outside of the box', but a helmet made from coconuts?*

16

The helmet is actually lined with foam which, in a crash, absorbs impact energy, while the coconut forms the protective shell.

One advantage is that the materials are actually lighter than those used in conventional helmets.

# CRINOLINE

They may be synonymous with starchy, repressive 19th-century mores, but crinoline dresses were once the must-have fashion statement for women of all social levels. Dress fabrics used around the time were extremely heavy and required several uncomfortable layers of petticoats to support them. But the introduction of light, metal-hooped crinoline cages freed the wearer from the tyranny of tangling petticoats. The steel of the cage was light and springy enough to bend and then spring back into shape if, for example, the lady passed through a narrow doorway.

However, diameters kept on getting wider, reaching a peak of almost 180cm (6ft) and, by 1860, the 'bird-cage contraptions' were mercilessly ridiculed by satirists. Furthermore, they were often the cause of social embarrassment. If the lady sat down without properly spreading her skirts first, the cage could spring the dress up in her face. They also swung about in the wind, lifting the dress to reveal flashes of leg or underwear. And ladies had to endure heavy, boned corsets, which needed to be worn to support the crinoline. By the end of the 1860s dressmakers turned to more streamlined styles.

*'Also provides useful protection for potted plants from frost.'*

Left: *Crinoline dresses were an advance on what had preceded them in terms of comfort, but they were still extremely restrictive to wear, and were often the cause of social embarrassment. A satirical magazine named the craze 'Crinolinemania'.*

It was said that the average-sized room could only fit two or three women wearing crinolines.

While the hoops seemed cumbersome, they were actually made of light metal that bent when passing through a narrow doorway or carriage opening.

19

# ELECTRIC HAIRBRUSH

If ever there was a man who added to the dire reputation of 19th-century quacks, it was Doctor Scott. There was scarcely a disease known to man that could not, so he boasted, be cured by his electric hair brushes. Anything from rheumatics to malaria to constipation would be shown the door by a few robust sweeps of the scalp. His brushes weren't even electric, but actually relying on magnetized iron rods in their handles. But the good doctor never let a trifle like that get in the way. 'There need not be a sick person in America…if our appliances became a part of the wardrobe of every lady and gentleman' ran one of his adverts. For a dozen years or so, it seems, people believed him.

*'Will positively produce a rapid growth of hair on bald heads.'*

Left: *Doctor Scott was so unscrupulous that he even claimed the brushes lost their healing powers if used by more than one person. It obviously meant bigger sales if every member of the family had to buy their own.*

Getting the products patented was a shrewd strategy. He could then mention the fact in the ads, making it appear as if they had an official stamp of approval for the curative claims.

Doctor Scott did generally steer clear of making any medical claims for his devices when he actually made his patent applications. But when it came to advertising, he clearly laid it on thick and heavy.

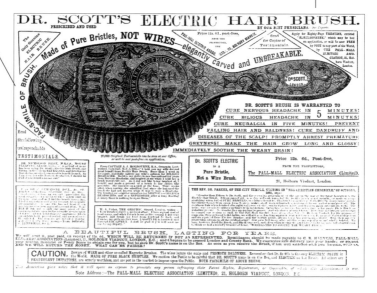

# EXPANDING SHOES

A rapidly growing child can place a huge stress on the parental budget, with clothes and shoes quickly being outgrown. Footwear is expensive, an issue that the expandable pair of shoes aims to address. At the push of a button on the heel, the shoe can be set to three different sizes. An adjusting mechanism locks it in so that the child is not exposed to loose or tight-fitting footwear. The idea of expanding shoes could certainly be good for an initial laugh or two in the playground, but once word gets around that these are a cheap option for cash-strapped parents, the attraction will soon wear off.

**RATINGS**

**STYLE COUNT:** ★★☆☆☆
**ORIGINALITY:** ★★★☆☆
**USEFULNESS:** ★★★★☆
**NERD APPEAL:** ☆☆☆☆☆
**LONGEVITY:** Despite a few child protests, sensitivity to changes in footwear fashion could see this one last.

*'The shoe that grew!'*

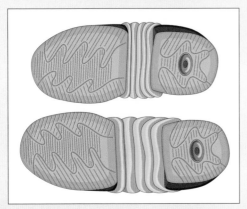

**Left:** *The button on the heel allows the shoe to be adjusted in half size increments up to a full length.*

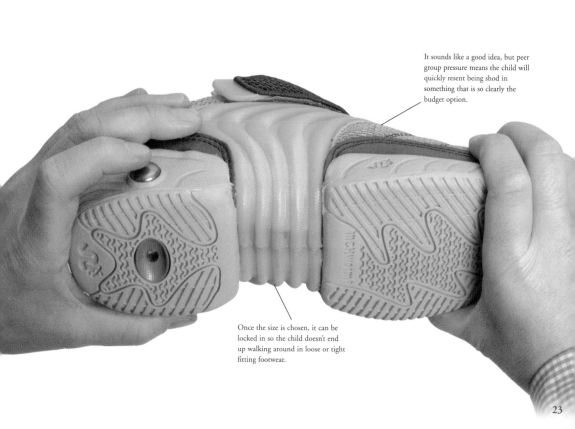

It sounds like a good idea, but peer group pressure means the child will quickly resent being shod in something that is so clearly the budget option.

Once the size is chosen, it can be locked in so the child doesn't end up walking around in loose or tight fitting footwear.

23

# EXTENDING NAIL CLIPPERS

You know you have reached the point of infirmity when you can no longer reach down to clip your toenails. It's at this point in your life that you must accept the necessity of a trusty new ally, the extending nail clipper.

Suddenly the task becomes as easy as picking cherries off a tree. There are clippers with angled heads to reduce the stress on wrists and hands, and others with cushioned palm and finger rests for a more comfortable grip (no more turning the air blue as you fling the clippers across the bedroom floor in frustration). One thoughtful manufacturer has designed some that you can sling in the dishwasher for cleaning without jamming up the mechanism, and another has brought one out with a magnifying glass attached, so you can actually see which nail you are cutting. Old age? Bring it on!

*'Using them's easy when you toe how.'*

Left: *It's easy to laugh but, of course, it's not just the elderly who might need them. Sufferers of arthritis or chronic back problems, the clinically obese, or pregnant women could find them very useful.*

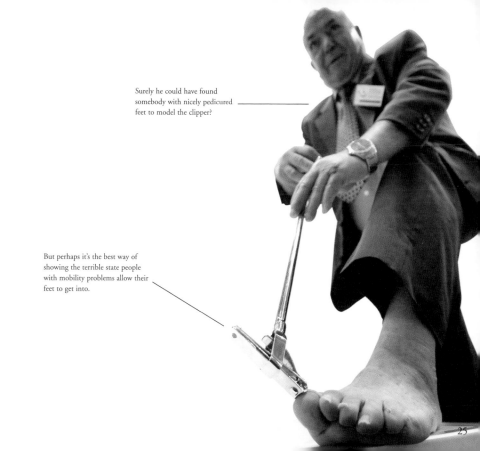

Surely he could have found somebody with nicely pedicured feet to model the clipper?

But perhaps it's the best way of showing the terrible state people with mobility problems allow their feet to get into.

# FAUCET UMBRELLA

This umbrella is essentially a water-collection device, a customized version of a conventional umbrella transformed into an item to help save water resources. Why simply allow rainwater to splash uselessly over your umbrella, or soak your clothes, when you could be using the opportunity to be store it for future use? All you need to do is pick up a funnel, feed it through a hole in the top of the umbrella, with a tube for the

water to run down. The water then drains out of an outlet you've sensibly substituted for the handle, into an attached bottle. You won't bring an end to droughts or the vicissitudes of climate change, but at least you can fool yourself into thinking you are.

*'No walk in the rain is a waste of time when you've got a faucet umbrella.'*

Left: *They say the simplest ideas are best, but the faucet umbrella does little to support that supposition.*

This funnel could be put to far better use in the kitchen.

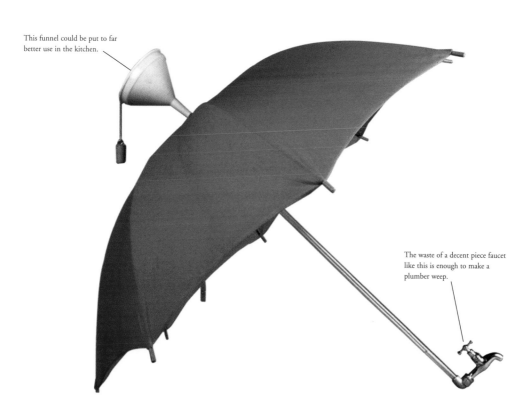

The waste of a decent piece faucet like this is enough to make a plumber weep.

27

# HEAD UMBRELLA

There are some things that were designed to be funny that aren't, and simply leave you wanting to thump someone. But there other items, also knowingly designed and worn to raise a laugh, that succeed. Despite being so blatant, they still touch the human funny bone.

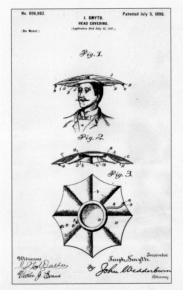

The head umbrella has no pretences to being anything other than ridiculous, and it is. But has it any practical use at all? Certainly the elastic one-size-fits-all headband is a plus. It keeps the sun off, so is great for picnics or a walk along the beach. In a shower, it also frees up the hands for zimmer-frame shufflers or golfers or fishermen. It's simply amazing that something so earnestly patented as long ago as 1898 remains a handy all-purpose gadget even today.

*'Great for a round of golf in the rain.'*

Left: *In the 1960s, it was fashionable for office workers to don head umbrellas for lunchtime reading sessions in the park.*

The basic idea for a head umbrella hasn't changed much in one hundred years.

The one time you won't laugh at someone wearing a head umbrella is if they're sitting in front of you at a sporting event.

# INFLATABLE BRA

Eighteenth-century women enhanced their figures with 'Bosom Friend' quilted pads. Even in the chaste 1850s, 'Lemon Bosoms' were available to those wishing to fill out their busts. But it was the desire to emulate 1950s stars like Jayne Mansfield and Marilyn Monroe that led to the brief explosion onto the scene of the inflatable bra. These took spring-loaded cups and foam-rubber falsies to the next level. By blowing into a tube, the bra cups were individually inflated. Unfortunately they could just as easily burst, and a female taking a ride in an airplane ran the high risk of one or both of the cups actually inflating even further, before exploding with a loud bang, to the amusement of fellow passengers.

*'What goes up must come down.'*

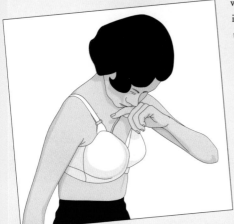

Left: *Just put your lips together – and blow. Preferably in private though.*

30

Many a 1950s girl envied the figure of Marilyn Monroe, though the thinking behind the inflatable bra was probably as much about male fantasy.

The fashion for tight sweaters meant there was little chance of hiding a sudden decompression.

31

# MAKEUP EYEGLASSES

As if getting older was not bad enough, magnifying makeup eyeglasses ram the point home a little further for the woman of a certain age by suggesting she probably needs a helping hand applying makeup. More than thirty million American females past the age of forty encounter difficulty applying their makeup because of blurry vision, goes the advertising blurb. Stop slapping on those wanton layers of mascara that make you look like Coco the Clown, you blind old fool, it cruelly implies, and while you're at it, say goodbye to wonky eyebrows, too.

*'Also worn by ladies after that retro late 70s look.'*

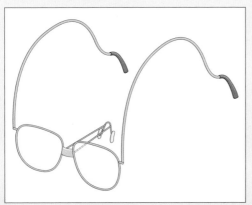

**Left:** *It's downhill all the way for women over forty, the advertisers of magnifying makeup glasses are keen to impress upon us. Whatever happened to the idea that life actually begins at forty?*

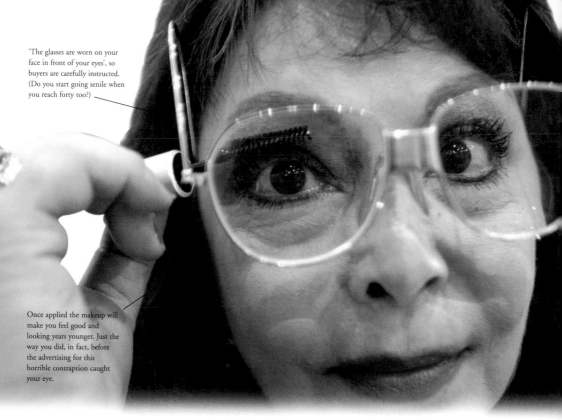

'The glasses are worn on your face in front of your eyes', so buyers are carefully instructed. (Do you start going senile when you reach forty too?)

Once applied the makeup will make you feel good and looking years younger. Just the way you did, in fact, before the advertising for this horrible contraption caught your eye.

33

# MOUSTACHE PROTECTOR

Eating and drinking has always presented something of a challenge for men with large amounts of facial hair. However, in the Victorian and Edwardian eras, when a bristling moustache and whiskers were a gentleman's badge of masculinity, the potential for social embarrassment must have been considerable. Imagine the harrumphing indignity caused when a morsel of food got stuck on a side bar of the 'tache, or soup formed an unsightly stain.

Step forward the Acme Novelty Company from Nebraska with their

moustache guard. It was made of gold and silver plate and, attached to a piece of elastic, was said to preclude the need for napkins. Unfortunately, anyone donning it risked even greater embarrassment, because it made them look not merely ridiculous, but positively sinister.

**RATINGS:**
**STYLE COUNT:** ★★★★★
**ORIGINALITY:** ★★★★☆
**USEFULNESS:** ★★★★☆
**NERD APPEAL:** ★★★★☆
**LONGEVITY:** Unlikely to have had the manufacturers of napkins fearing for their future.

*'Makes you look like the Phantom of the Opera.'*

**Left:** *Apart from looking silly, wearing the moustache guard, even in its larger size, was unlikely to be much use, given the fashion for full-blooded moustaches in the 19th century.*

Appearance and table
manners meant everything
in 19th and much of the
20th centuries, but the
fashion for voluminous
amounts of facial hair didn't
make life easy for the
gentleman dining at his
club or in mixed company.

Another idea was to attach a piece of
metal with a hole in it to a teacup
and spoon. This enabled 'gentlemen'
to sip tea or soup without spoiling
their precious moustaches.

# MULLET HAIRCUT

One of the more polite terms coined for a mullet was the ape drape. Essentially a phenomenon of the 1980s and early 1990s, it must rank as one of the ugliest hair stylings in the history of civilization. In fact, 'civilization' is probably the wrong word, because these days it is generally only to be seen on redneck types. For a nightmarishly long period of time, however, it was as if every major male sporting figure or popular entertainer

was wearing one. The mullet at its most reprehensible was a kind of crew-cut affair at the front, while the back of the hair was left long. These straggly locks were often highlighted. There was also a bouffant mullet and other variations along half-hearted rat's tail lines. All shared a defining characteristic – they looked absolutely ridiculous.

## RATINGS

**STYLE COUNT:** ★★★★★
**ORIGINALITY:** ★★★★★ Some say you can detect that the Sphinx is wearing a mullet. No wonder the dynasty of the pharaohs died out.
**USEFULNESS:** ★★★★★
**NERD APPEAL:** ★★★★★
**LONGEVITY:** Whisper it quietly, but an 'ironic' retro 80s look is very much on the way back.

### 'Business in the front, party in the back.'

**Left:** *From the safety of the 21st century, you can look back and laugh almost affectionately at the stereotypical image of the mulleted male, but for a time he was horribly real.*

The curious thing with this mullet, sported by Billy Ray (father of Miley) Cyrus – is the attempt to combine macho hunkiness with effeminate highlights.

There is a theory that the mullet was created in a German police station in 1980. A hairdresser had had her bag snatched and went to report it. In flicking through the police IdentiKit book to draw up a picture of the thief, the hairdresser chanced upon the combination of a cropped head of hair on top with the lower jaw and shoulder-length hair of another man. No such suspect fitted that description and she never found her bag, but she went on to create her own criminal haircut.

# NOSE SHAPER

Had sales of Mr Trilety's Model 25 Nose Shaper truly taken off, who knows how much money would have been saved on plastic surgery down the years? Back in the 1920s, his marketing pitch was that, for a small fee, his device would reshape the nose by remodelling the cartilage and fleshy parts 'quickly, safely and painlessly'. You can still pick up similar gadgets today that, for just 20 minutes' wear, claim the same results. But, like the

one sold by Mr Trilety, they probably won't work, unless you are young and have a skin and bone structure that is still developing.

HOW TO OBTAIN
**A Perfect Looking Nose**

My latest improved Model 25 corrects now ill-shaped noses quickly, painlessly, permanently and comfortably at home. It is the only nose-shaping appliance of precise adjustment and a safe and guaranteed patent device that will actually give you a perfect looking nose. Write for free bb which tells you how to obtain a perfect looking nose. M. Trilety, Face Noseshaping Specialist, Dept. 2414, Binghamton, N.Y.

RESHAPE your NOSE to beautiful proportions — while you sleep!

ANITA NOSE ADJUSTER is SAFE, painless, comfortable. Speedy, permanent results guaranteed. Doctors praise it. No metal to harm you. Small cost. Write for FREE BOOKLET

Gold Medal Won 1923

ANITA CO. 173 ANITA Bldg. NEWARK, N. J.

| RATINGS | |
|---|---|
| **STYLE COUNT:** | ★★★★★ |
| **ORIGINALITY:** | ★★★★★ |
| **USEFULNESS:** | ★★★★★ |
| **NERD APPEAL:** | ★★★★★ |
| **LONGEVITY:** | Search the internet and variants on the Model 25 are not hard to find. |

*'Great for perfecting that Hannibal Lecter grin.'*

**Left:** *Interestingly, his marketing was based around the idea that 'in this age' making the most of your appearance was an absolute necessity because the world would judge you mainly on your looks. Not much changes.*

The difference between Trilety's and later nose shapers was that his had to be worn continuously, so you'd have to be very dissatisfied with your appearance to take him up on his offer.

Modern nose shapers don't look much different from those that were on the market 80 years ago. They are probably no more effective either.

# PNEUMATIC ATTACHMENT FOR TROUSERS

At the right-hand corner of Mone Isaacs's diagram for a pneumatic attachment for trousers is an unintelligible scrawl of handwriting. It's clear that the U.S. Patents Office attorney who had to oversee the application was laughing so much, he couldn't see to write his name. More worrying is Mone's signature – bolt upright and clear – above it. Here was a man for whom pneumatic trouser attachments were a deadly serious matter.

His idea was that the trousers (adapted for either sex, he hastened to explain) could be pneumatically inflated in the crotch area, and he carefully provided front and back 'elevations' of the said garment, though not, sadly, with a person inside.

The inflatable attachment was a cushion, or 'reservoir', blown up by a bellows-like pump or tube, which Mone argued made for a resilient seat 'especially serviceable for bicycle riders'. While in the 1890s, uncomfortable 'bone-shaker' bikes were still much to be seen, reports of pneumatic trouser-wearing riders have proved untraceable.

(No Model)
M. R. ISAACS.
PNEUMATIC ATTACHMENT FOR TROUSERS.
No. 585,210.          Patented June 29, 1897.

Fig. 1.

*'Useful for hernia sufferers.'*

Left: *Late 19th-century bicycle riding could be an uncomfortable experience. But natural human pride means most would have baulked at softening the journey in trousers with an artificially inflated crotch.*

40

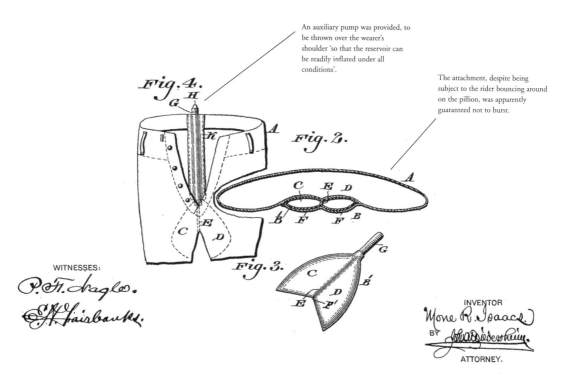

An auxiliary pump was provided, to be thrown over the wearer's shoulder 'so that the reservoir can be readily inflated under all conditions'.

The attachment, despite being subject to the rider bouncing around on the pillion, was apparently guaranteed not to burst.

Fig. 4.

Fig. 2.

Fig. 3.

WITNESSES:

P. F. Eagle.

E. H. Fairbanks.

INVENTOR
Mone R. Isaacs
BY
Attorney.

41

# RAIN GLASSES/GOGGLES

A frequent complaint of wearers of glasses is the nuisance of having to constantly wipe the lenses of rain or snow when caught in adverse weather conditions. It's bad enough for the average wearer, but imagine what a perpetual aggravation it must be for outdoor workers. It's a surprise, therefore, that the really smart idea of a clip-on, motorized 'windshield-wiper' attachment never caught on. This handy, detachable gadget came with a tiny two-speed motor, powered by ordinary batteries, which sent two little wiper blades across the lenses to clear them. There were even contingency plans to adapt them for use on motorcycle helmet face shields.

*'You'll make a real a spectacle of yourself.'*

Left: *Note very useful balancing device beneath the chin, adding a satisfyingly symmetrical element.*

42

Perhaps the marketing strategy wasn't right. Using a smiley little brunette to model them rather than a grinning weirdo would have been a smarter option.

It's a mystery how rain glasses never quite took off. Not only were they extremely practical, but as this dude shows us, they are a match for many an off-the-wall optical fashion accessory.

# REFRIGERATING SUIT (APPARATUS)

Ideas about inventing refrigerated, or cooling, suits seem to have exercised the minds of technicians for decades, but no one can have come up with a more spaced-out contraption than R.S. Gaugler did in 1937 with his dual purpose 'refrigerating apparatus'.

This was a refrigeration unit in the shape of a small fridge attached to a specially designed suit and blanket. Cooling air was pumped into the suit or blankets along tubes inserted into the fabric.

The suit would have been incommodious and desperately impractical, with the connecting lead too short to render it any use for performing tasks requiring movement. Meanwhile the drone of the refrigerating unit would have ensured that the hope of getting any sleep under your cool – but incredibly heavy – blanket was nonexistent.

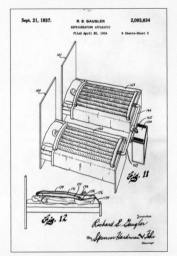

Sept. 21, 1937.   R. S. GAUGLER   2,093,834
REFRIGERATING APPARATUS
Filed April 30, 1934   6 Sheets-Sheet 5

*Fig. 11*

*Fig. 12*

Richard S. Gaugler

*by* Spencer Hardman & Fihe

*'Perfect for that Saharan journey you always wanted to make...'*

Left: *However idiotic Gaugler's invention looks, it was a step on the way to the cooling vests worn today by soldiers in hot climates, or emergency workers combating major fires.*

44

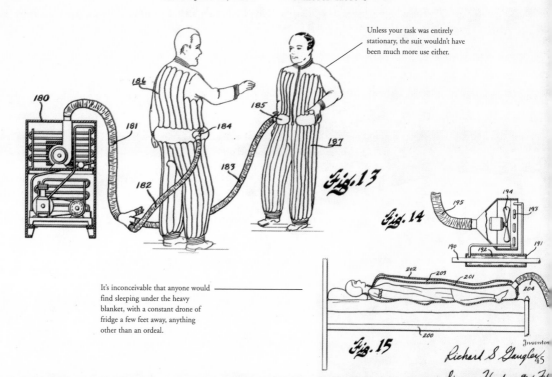

Sept. 21, 1937.

R. S. GAUGLER

2,093,834

REFRIGERATING APPARATUS

Filed April 30, 1934

6 Sheets—Sheet 6

Unless your task was entirely stationary, the suit wouldn't have been much more use either.

*Fig. 13*

*Fig. 14*

It's inconceivable that anyone would find sleeping under the heavy blanket, with a constant drone of fridge a few feet away, anything other than an ordeal.

*Fig. 15*

Inventor

Richard S Gaugler

# SLEEPING JACKET

Going anywhere on buses and trains can literally result in a pain in the neck if you are a habitual dozer. The sleeping jacket is specially designed to rectify the problem, however, working on research that has shown that it is actually possible to sleep in a rigid, upright state. A big hefty collar keeps your neck from rolling around, the lapels serve as pillows, and the sides tighten to keep your core posture upright. Unravel the cuffs and

they are transformed into nice warm mittens. Of course, you'll get lots of stares from fellow passengers, because the collar in particular looks pretty unsightly. But the invention was sponsored by a designer of vacuum cleaners, so what can you expect?

*'Line the pockets with old newspapers and you'll have the perfect tramp on the subway look.'*

Left: *Everyone knows the feeling of dozing off on a train and waking to find yourself with an uncomfortable crick in the neck. But how far are you prepared to compromise concerns about appearance in search for comfort?*

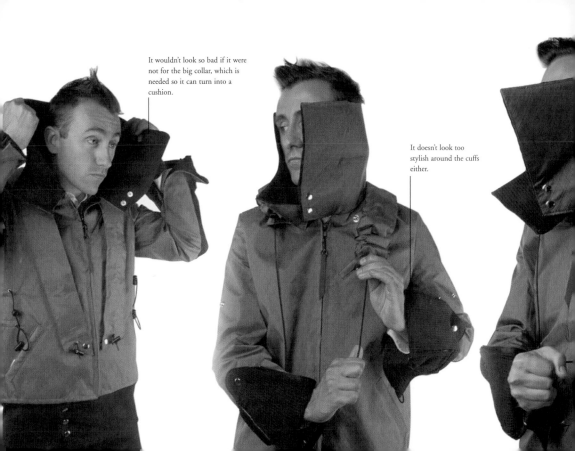

It wouldn't look so bad if it were not for the big collar, which is needed so it can turn into a cushion.

It doesn't look too stylish around the cuffs either.

# STEALTH UMBRELLA

The makers of the Stealth umbrella have taken the idea of doing battle with the elements a step further. By using aerodynamic technology used in fighter bombers, they claim their umbrella is less prone to blowing inside out than conventional round ones. The trouble is, being shorter at the front, it fails to afford sufficient protection in a heavy downpour, so you get drenched anyway. Because wind is devious and likes to attack from different directions, you find it whipping around and blowing the umbrella inside out. The maker says it has accounted for this with a handle that, if grasped lightly, simply swivels around to the right direction. Try it, though, and you risk the whole umbrella being carried away in the gale. But you have to admit, Stealths do look chic.

*'You'll never have to worry about radar detection.'*

Left: *The Stealths are expensively priced, so you'd expect them to look a little more stylish than regular cheap foldables – and they do.*

The canopy doesn't have the usual nasty, antisocial spokes. But fellow pedestrians may still give you a wide berth when they see you approaching with this oddly shaped contraption.

The short front of the canopy means donning a full length raincoat is advisable.

49

# TWO-WAY SHOE

If you've ever experienced that momentary frisson of irritation when having to switch around a pair of shoes you've left by the doormat to put them on, you'll recognize the philosophy behind the two-way shoe. When entering the house, you can simply slide off your shoes, content in the knowledge that when you want to go out again, you'll be able to slip back into them without the bother of having to turn them around.

The shoe actually originates from the Japanese Chindogu movement, or Art of Useless Inventions. The idea is that the inventions 'work', but have no practical use. On that score the two-way shoe certainly earns full marks.

**RATINGS**

STYLE COUNT: ★★★★★
ORIGINALITY: ★★★★★
USEFULNESS: ★★★★★
NERD APPEAL: ★★★★★
LONGEVITY: If there is any hope of longevity, it lies in the statement as much as the item.

*'For the person who's not sure whether they are coming or going.'*

Left: *Chindogu inventions cannot be patented, or profited from (not that there was any chance of it happening anyway).*

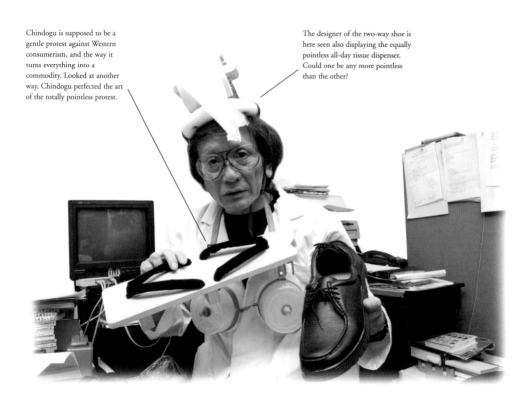

Chindogu is supposed to be a gentle protest against Western consumerism, and the way it turns everything into a commodity. Looked at another way, Chindogu perfected the art of the totally pointless protest.

The designer of the two-way shoe is here seen also displaying the equally pointless all-day tissue dispenser. Could one be any more pointless than the other?

# VENETIAN BLIND SUNGLASSES

Although the heyday of Venetian blind sunglasses is generally taken to be the 1980s, there are images showing people wearing them in the 1950s. In recent years, they have also enjoyed something of a comeback. But their longevity, or bounce-back-ability, still doesn't make them any more notable. They are simply a useless fashion accessory for those with no taste. Generally speaking, they make no pretence of protecting against UV rays. In most cases, they don't have any lenses in them at all, and merely consist of a series of plastic horizontal bars across the frames. There are many people who love them, and as many who hate them. For most people, however, these glasses remain a matter of total indifference.

**RATINGS**

| | |
|---|---|
| **STYLE COUNT:** | ★☆☆☆☆ |
| **ORIGINALITY:** | ★★☆☆☆ |
| **USEFULNESS:** | ☆☆☆☆☆ |
| **NERD APPEAL:** | ★★★★★ |
| **LONGEVITY:** | Until the latest craze is replaced by something equally vacuous. |

### 'The Venetian Blinds that don't work.'

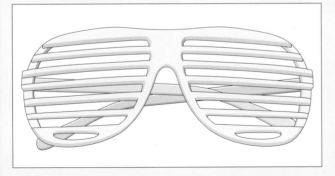

Left: *Unashamedly a fashion accessory, to the unconverted, they just look tacky, but they are certainly not cheap.*

One thing you can guarantee about these tasteless items is that they are available in an array of revolting 'shades'.

They've become a big fashion statement again lately, but in a few years time, we'll look back and laugh at how ridiculous they looked.

53

# WEAR-ME UMBRELLA

The wearable umbrella might well be the biggest innovation in umbrella technology since its invention 3000 years ago. This is an umbrella you actually clip on and wear rather than hoist above your head. There are no poles, points or rods and, covering your upper body, it doesn't wobble all over the place. It's claimed that it cannot blow inside out, and the big plus is that both hands are free for bag carrying. The plastic canopy is also fully transparent – allowing a full view of all those nervous glances you'll be attracting.

**RATINGS**

**STYLE COUNT:** ★★★☆☆ Quite stylish in a futuristic kind of way

**ORIGINALITY:** ★★★★☆

**USEFULNESS:** ★★★★☆

**NERD APPEAL:** ★★☆☆☆

**LONGEVITY:** The only thing that's likely to work against them is their outlandish look.

*'Look! No hands!'*

Left: *Early wear-me umbrellas were more on 'jokey' lines. This one looks like she's simply recycled a cinema usherette's ice cream tray.*

New wear-me umbrellas strap to your shoulders and cover your upper body, leaving your hands free. They open and close by pushing a button.

People say new wear-me umbrellas look silly, but in the middle of a rain storm, who's going to be looking?

55

# ZIG-ZAG COMB

A late noughties trend for females was to have their hair done with a zig-zag parting. The effect was not particularly easy to achieve, however, and was best left to stylists in the hair salons. That was until some genius came up with the zig-zag comb. It was slickly designed, with a contoured handle for an easy grip. Three cylindrical stainless steel pins in a triangular shape created the all-important parting, as the comb was moved from back to front in a zig-zag motion. Each comb came with careful instructions on how to use it. Before long every young girl seemed to have a zig-zag, to the extent that a walk down the street left you feeling completely giddy.

**RATINGS**
STYLE COUNT: ★★★☆☆
ORIGINALITY: ★★★☆☆
USEFULNESS: ★★☆☆☆
NERD APPEAL: ☆☆☆☆☆
LONGEVITY: Vulnerable to being tossed away on the flotsam and jetsam of female fashion.

*'Hair today, gone tomorrow.'*

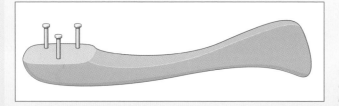

Left: *Thanks to a 'click-in' head, you could use the zig–zag comb in your left or your right hand, whichever you fancied.*

It was crucial to get a perfectly balanced zig-zag, otherwise you looked like the victim of a bad hair transplant.

Like the best of fashions, the zig-zag haircut soon looked very dated.

# KITCHEN CREATIONS

You have to wonder how the Stone Age housewife coped. Possessing only a bit of old flint and a chipped earthenware bowl, she was expected to rustle up a hearty meal in under the hour when her hungry, hunter-gatherer husband walked in with an elk on his back, and slung it in the pot. Now we get down to work with microwave ovens and electric carving knives and everyone's happy. But alongside the quest for efficiency and speed in the kitchen has been an onslaught of pointless little add-ons that aid no one, from triangular teabags (a complete con aimed squarely at the 'more money than sense brigade') to illuminated ice cubes, which were strictly childrens' birthday party fare. Consider, too, the egg-o-matic, which was really just an excuse for its lazy proprietor to spend a couple more hours in front of the television, though it was convenient for users who found themselves short of an egg when the stores were closed. Only the pop-top can is really worth three cheers, and even that has questions to answer about its contribution to the proliferation of litter in the streets.

Left: *You were always sure to draw a crowd when you visited an Egg-O-Matic machine.*

# *AUTOMATIC EGG COOKER*

Eggs may be a dietary staple, but their fragility and difficulty to cook just right would try the patience of Job. The automatic egg cooker promises to remedy all that, except that it brings a wagon load of baggage of its own. It takes you months of experimentation to find the right setting: at first the timer goes off straight away without having even begun to cook the egg, then it doesn't go off at all. Or maybe it cooks the egg perfectly, but then you can't

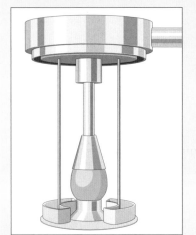

get the lid off because it's too hot. Or the tray the egg is supposed to sit in is too small, so the egg slips out, breaks and the yoke spills out and makes a dreadful mess. Maybe using a good old saucepan and boiling up some water wasn't such a bad idea after all.

## RATINGS

**STYLE COUNT:** ★★☆☆☆
**ORIGINALITY:** ★★★☆☆
**USEFULNESS:** ★★★☆☆
**NERD APPEAL:** ☆☆☆☆☆
**LONGEVITY:** Our hunger for convenience in the kitchen means this invention is here to stay.

### *'Bound to make Humpty Dumpty weep.'*

Left: *It's so easy to stick an egg in a saucepan of water and walk away and forget about it that an automatic cooker seems a sure-fire improvement. But this Walter Mitty-style attitude to the gadgets has time and again been proven wrong.*

Set the dial, sit back and wait, the instructions tell you. If only it were that simple!

Hard boiled or poached, experience proves there's no shortcut to cooking an egg.

61

# BUCKLE'S DUNKSAFE

Some great minds have spent hours sweating over a steaming cup of tea, trying to come to terms with the science of cookie dunking. The latest, peer reviewed, research suggests that the bigger the cookie diameter, the less dunking time you've got before bits break off and sink soggily to the bottom of the cup. And it was to avert this calamity that Buckle's Dunksafe was devised. Nothing less than a self-contained unit with holes at the bottom, you simply place the cookie of your desire into the unit and the tea flows

through to avoid any spillage or crumbling.

Quite how you are supposed to drink the tea is another matter, but for dunking scientists, it has been hailed as a substantial breakthrough.

*'Worth celebrating with a national dunking day.'*

Left: *Whether on a desk carrying a cup of tea or coffee, many a frustrated fist has been banged over the soggy cookie problem.*

Research has shown that the taste of the cookie is enhanced ten times by dunking as opposed to eating it dry. So the eponymous inventor of Buckle's dunksafe could be on to a good thing.

But what if you like sipping your tea and dunking at the same time?

# COFFEE-FLAVOURED MINTS

Coffee drinkers aware of the pungent aromatic effects upon their breath of an early morning cup will take some convincing. However, extracts from the product could soon become a new, entirely unexpected entrant onto the market of breath mints and mouthwash.

The halitosis-effect is caused by the dehydrating impact of coffee, made more potent by its mixing with milk, and causing fermentation. However, scientists monitoring the actual coffee beans have found that they contain compounds that can prevent bacteria from releasing the gases behind halitosis.

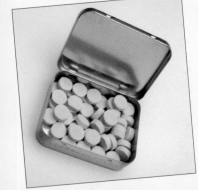

While the scientists have yet to identify which specific compounds contribute to the beneficial effects, the expectation is that future studies will be successful in determining the active ingredient, and that this can lead to the production of new 'bad breath' mints or pellets, or even chewing gum.

**RATINGS**

STYLE COUNT: ★★★★★
ORIGINALITY: ★★★★★
USEFULNESS: ★★★★★
NERD APPEAL: ★☆☆☆☆
LONGEVITY: Coffee drinkers, or those who've sat inhaling the noxious breath of coffee-drinking colleagues, may look askance. And who'd really want to suck on a mint that tastes like coffee?

*'Banish bad coffee breath by sucking a coffee-flavoured mint!'*

Left: *A cup of coffee in the morning may give you an energy lift but it can also leave you with halitosis.*

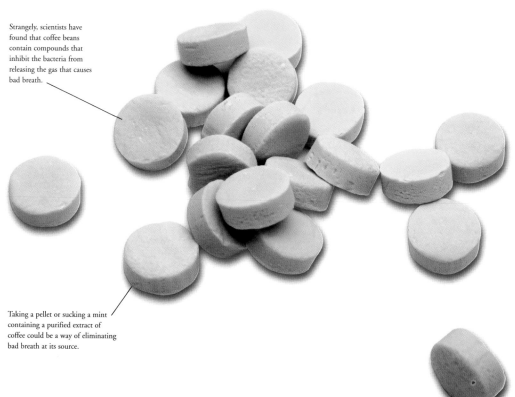

Strangely, scientists have found that coffee beans contain compounds that inhibit the bacteria from releasing the gas that causes bad breath.

Taking a pellet or sucking a mint containing a purified extract of coffee could be a way of eliminating bad breath at its source.

# DONUT DUNKER

You could tell Russell H. Oakes was a 'funny' inventor in the ha-ha rather than oddball sense by the way he'd sign off his ideas with the signature Professor 'Ratzin de Garrett' Oakes. It wouldn't make you laugh much now, but in the 1930s he apparently had Midwest audiences rolling in the aisles with demonstrations of his wacky ideas.

When he turned up with something aimed at 'helping' habitual consumers of that hallowed item of American hedonism, the donut, he was guaranteed an enthusiastic reception.

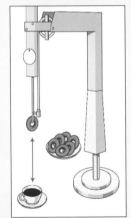

Donut dunkers, he explained, were forever scalding their fingers, or dripping coffee on crisp, clean tablecloths, a problem his dripless donut dunker could solve. This unit, resting on castors, had a long arm that held the donut, and swung across to dunk it in the coffee. As the person raised the donut to their mouth, a cup or pan swung over to catch any drops. How the audience roared and clapped – and nodded to each other in amazement at the way the good professor managed to keep coming up with these wild and crazy gadgets.

*'Saves a fortune in tablecloth laundry bills.'*

Left: *Americans have drooled over donuts for decades. In 1934, they were billed as 'the hit food of the Century of Progress,' but no one has ever truly come up with a solution to the problem of the drips.*

Oakes's dripless donut dunker was only one attempt at answering the problem. Pulley-style donut-dunking contraptions were another gimmick found in coffee shops.

That spotlessly white shirt is guaranteed to remain that way thanks to the dripless donut dunker. How much longer it will fit him, if his donut consumption habit continues, is another question.

67

# EGG-O-MAT

On weekdays in the late 1940s, New Jersey farmer Camillo Epstein and his wife used to sell eggs on a stand in front of their house. To give themselves a break in the evenings and on the weekends, however, Camillo concocted the Egg-O-Mat, a coin-operated 24-hour egg dispensing machine. The little wooden shed was refrigerated by an air-conditioning unit at the back.

It looked a bit ramshackle, but locals were eggstatic at Camillo's ingenuity, and it became a roadside fixture in their hometown of Warren, New Jersey, for over 40 years. Camillo took the issue of vandalism very seriously and, to deter youngsters who viewed it as some kind of amusement arcade machine, he installed an intercom system so he could be alerted of any suspicious activity back at his house. Mainly, though, it just relayed the happy buzz of contented locals, picking up eggs around the clock. Happy days.

*'Eggs of Distinction 24 Hours a Day.'*

Left: *The Egg-O-Matic: there were many variations on Camillo Epstein's Egg-O-Mat idea in the days before open-all-hours supermarkets.*

The success of the machines depended on the farmers keeping them stocked with fresh eggs. Otherwise they were just dispensing the 'oldest yolks in the book'.

The similarly named Egg-O-Matic was another egg-dispensing machine of the period, devised by Hector Garza.

EGG -O- MATIC

6 EGGS AND 3ᵈ FOR 2⁄6

2⁄6 2⁄-

PLEASE CLOSE FLAP

69

# ILLUMINATED ICE CUBE

High on the list of the most pointless objects ever invented comes the battery-powered, illuminated ice cube. In fact, these are not ice cubes at all in the real sense, but are comedy cubes. While they look real (except for the fact that they don't melt), they change colour every few moments. The gel they are filled with does actually help keep your drink cool but, other than as a novelty at children's birthday parties, they are unlikely to take the world of social get-togethers by storm.

## RATINGS
STYLE COUNT: ★★★★★
ORIGINALITY: ★★★★★
USEFULNESS: ★★★★★
NERD APPEAL: ★★★★★
LONGEVITY: Possibly as one for the kiddies.

*'So what's wrong with plain ice cubes?'*

Left: *Party novelties are not meant to be serious, but whoever came up with illuminated ice cubes clearly had too much time on their hands.*

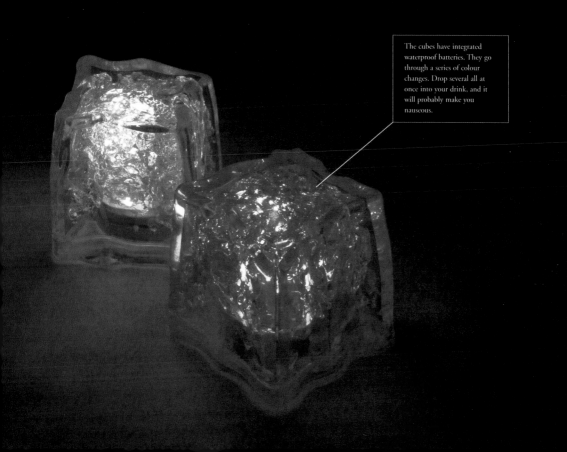

The cubes have integrated waterproof batteries. They go through a series of colour changes. Drop several all at once into your drink, and it will probably make you nauseous.

# POP-TOP CANS

Flat-top cans containing foods and beverages have been around since the early 1930s, but into the 1960s, you still couldn't access the contents without a can opener or 'churchkey'. Then Ohio toolmaker Ermal Fraze came up with the idea of a notched opening, built into the top of the can, that could be pierced and pushed open with a lever, or pull tab. By 1965, about three-quarters of all American brewing companies were using Fraze's

design, but problems persisted. Frequently, the tabs snapped off, leaving the can unopened and the hot, thirsty, would-be drinker boiling with frustration. The openings sometimes had sharp edges that caused cuts to lips. There were also a litter and environmental issues, because people threw away discarded lever rings which were injurious to the wildlife that injested them. However, by the late 1970s, the innovation of the 'stay-on' tab had helped resolve this issue.

*'Master the knack of administering exactly the right thumb pressure and you're in!'*

Left: *The pop-top can was a great idea (even if dismissed at the time) because it dispensed with the need to carry a can opener or 'churchkey'. But if the lever broke off, it was a cause of much frustration.*

Stay-on tabs were first developed in 1975, but they still can break off, so the problem hasn't entirely disappeared.

The early pull tab had a ring attached at the rivet which would come off and be thrown away, causing a contentious litter issue.

73

# SPAGHETTI FORK

It's generally agreed that it is unwise to order spaghetti on a dinner date or at a social meal with your boss unless you want to come across like someone who is a complete stranger to table manners. A lot of people therefore would be shouting for joy if someone came up with something like a reliable twirling spaghetti fork. All you'd have to do is turn up at the restaurant and, brandishing this little battery-operated beauty, fearlessly order your spaghetti. No more flicking spaghetti sauce into the eyes of fellow diners, or straining your wrist as you twist it 360 degrees while making undignified slurping noises. Unfortunately though, it wouldn't be total nirvana. Such a device would probably be very slow on the turns, and the battery would render it off-puttingly noisy.

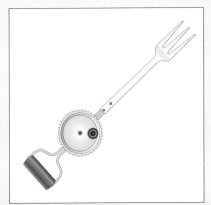

*'Ever wondered how the Italians managed for all these years?'*

Left: *The issue of how to eat spaghetti has exercised people's minds for decades, and, as shown here, consuming it in public has often been a form of social torture.*

For children, it's just a big laugh, but this old boy really should have known better.

The idea of utilizing other kitchen utensils to get around the problem can end up making you look even more ridiculous.

# *TRIANGULAR TEABAGS*

They sound like a not very cunning ruse by teamakers to add value to their products, but there are people who swear blind that the triangular tea bag really does make for a superior cup of tea. The bags are made of nylon or non-woven fabrics that allow the taste and aroma of the tea to diffuse more quickly and effectively. No big deal, you might think. Why not simply deploy the same materials on boring old round or square teabags without all the fuss? But here's the rub. Bags shaped like a triangle have more space than conventionally shaped ones. This means, apparently, that makers

can pack more tea leaves and buds into the bags, and this allows them to infuse better. But, funnily enough, more tea leaves and buds per bag also raises producer costs, which of course means you will also have to fork out a considerably larger amount of cash for a box of them than you would for regular teabags.

**RATINGS**

STYLE COUNT: ★★★☆☆ (They do look cool)

ORIGINALITY: ★★☆☆☆

USEFULNESS: ★★☆☆☆

NERD APPEAL: ☆☆☆☆☆

LONGEVITY: Smart advertising will ensure that as long as there are people willing to pay over the odds for average products, this one will always sell.

*'Specially priced for people with more money than sense.'*

Left: *One company actually sells triangular teabags that are individually handcrafted. But you still only get one cup out of them because, in the end, one teabag, whatever its shape, is much like another.*

The shape of the bag may be irrelevant but that extra bit of string and the tag sure are handy to avoid scalding your fingers.

The triangle shape is supposed to give more room for the tea leaves and buds to infuse.

# TV DINNER

TV dinners might stand as the perfect symbol of the quickening pace of postwar western society. Ideal for the time-poor individual, they are prepackaged meals, conveniently available frozen or chilled, ready to go straight in to the oven or microwave. However, they also represent the disconnection of the modern consumer from the process of food preparation. TV dinners make food into a commodity much like motor fuel, where the only matter of concern is whether it is unavailable on the store shelves or not. In addition, because of the degradation the food undergoes at processing stage, TV dinners – or ready meals – must be loaded with extra salts, fats, sugars and preservatives, which can make them deeply unhealthy.

*'Saves your time, but not your health.'*

Left: *TV dinners on a mass scale first became available in the 1950s. They arrived on the scene around the same time more people were acquiring televisions and eating in front of them, hence the name TV dinner.*

Early TV dinners on offer did not extend far beyond fried chicken, frozen peas and sweet potatoes, but an enormous variety of choices are available today.

Recent concerns about nutrition have led to a spate of Healthy Choice options arriving on supermarket shelves. However, they often taste just as bland – if not worse – than their sodium- and fat-laden predecessors.

# VEGETABLE PRESERVER

Modern vegetable keepers are neat, compact drawers that fit snugly in the fridge, where you needn't given them a second thought. Because they have ionic technology, manufacturers claim the vegetable keepers even reduce the types of gas that cause the food to wilt.

Earlier generations resorted to more primitive measures, such as storage in a cool, dark place, or even burying food like cabbages in the ground. So the inventors of the electronic vegetable preserver probably thought they were being very clever when they came up with the idea of bombarding vegetables with electronic pulses to reduce the speed of decay. However, it is unlikely that long-suffering mid-twentieth century consumers, even as accustomed to cumbersome kitchen implements as they are, shared the excitement.

*'A helping hand in the kitchen from Dr Frankenstein!'*

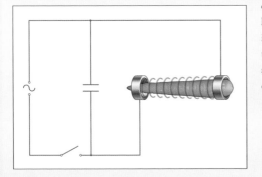

Left: Subjecting food products to pulsed electronic fields has long been the subject of much study as a means of food preservation. Introducing it for use in the kitchen has been more problematic.

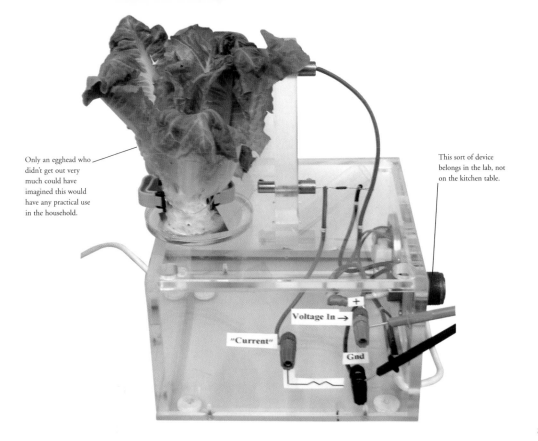

Only an egghead who didn't get out very much could have imagined this would have any practical use in the household.

This sort of device belongs in the lab, not on the kitchen table.

Voltage In →

"Current"

Gnd

How to Stop Killing Yourself

PETER J. STEINCROHN, M.D.

# HEALTH AND SAFETY GONE MAD

A lot of people get steamed up by what they understand to be the modern tyranny of health and safety regulations. The reality is that as lifestyles have become more diverse and complicated, and more sophisticated than times past, so potential perils to protect ourselves against have grown greater. It was only when Man began to explore life on the seabed that thoughts turned to such curious contraptions as marine lifesaving apparatus, for instance. There would have been no market for a buoyant swimsuit before family vacations beside the sea became popular in the last century.

Advanced medical knowledge of the connection between smoking and cancer lies behind the well meaning, if anodyne, electric cigarette, and an appreciation that we simply don't function as well when stressed behind the calm machine. One modern invention shows that there's still a place for a bit of old-fashioned quackery. Can anyone truly claim to have derived much benefit from a session with the age-defying tuning fork?

Left: *In the 1950s, you had more chance of perishing via a fireproof smoking device than succumbing to cancer.*

# ACUPUNCTURE BED

The bed of a thousand nails is a standard prop of an old-fashioned magician's routine. While it might look like death by a thousand cuts for the person foolhardy to try their luck on it – provided the nails are distributed in the correct way – the nature of body-weight distribution ensures that there is not enough pressure to break the skin. But therapeutic gains from lying on a bed of nails have been a traditional technique of acupuncture for even longer. The pressure points are said to stimulate the release of endorphins and healing energies, and enhance a sense of well-being. You can be forgiven, however, for remaining convinced that there must be a thousand easier ways to achieve similar results.

**RATINGS**

**STYLE COUNT:** ★★★★★
**ORIGINALITY:** ★★★★★
**USEFULNESS:** ★★★★★ Acupuncturists would claim it has its uses.
**NERD APPEAL:** ★★★★★
**LONGEVITY:** Thousands of years.

### 'We've got you nailed!'

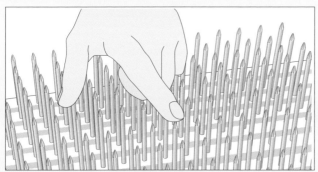

Left: *The acupuncture bed has a long tradition in eastern healing remedies. But those nails still look distinctly uninviting.*

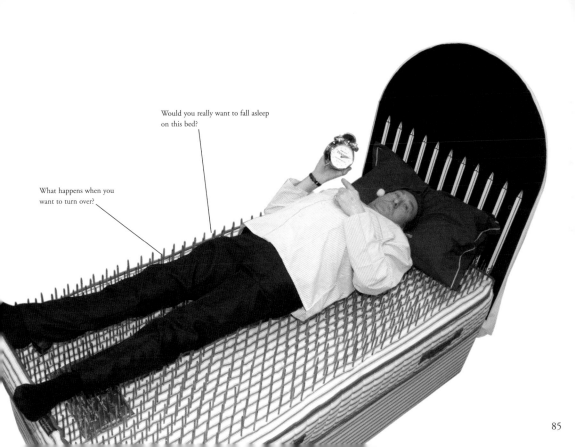

Would you really want to fall asleep on this bed?

What happens when you want to turn over?

85

# ALL-DAY TISSUE DISPENSER

There's nothing more miserable than an acute attack of hay fever, but the All-Day Tissue Dispenser takes remedies for the problem to levels of the absurd. What better way to deal with the runny eyes and nose that require constant recourse to hankies or tissues, than to have a roll of toilet paper

handily perched on your head? You need never reach down into your pocket for that increasingly sodden, useless piece of tissue again. Of course, the All-Day Tissue Dispenser is an 'invention' only in the most surreal sense. Silly, pointless, but, in a warped sort of way, quite cool.

*'You'll look great. Swollen eyelids, a big red nose, and a piece of toilet paper flapping down over your forehead!'*

Left: *Wearing the roll sideways is probably a smarter option. That way, it doesn't stop you from seeing where you are going.*

Non hay-fever sufferers probably don't appreciate how rotten the allergy makes people feel, so perhaps this is one way of making your utter misery clear.

Have you noticed how it always seems to be the Japanese who come up with these things?

87

# AGE-DEFYING TUNING FORK

On the quackometer scale, tuning forks used for tuning sagging or maturing skin are probably up there with magic crystal balls. Its supposed links to ancient Chinese medicine and acupuncture might arouse your curiosity but when you learn that the forks are also calibrated to the frequencies of the sun, moon and several planets, alarm bells start going off in your head. The metal forks come in various sizes, each one deployed for their unique quality of vibration. They are said to release healing energies on areas difficult to treat through acupuncture, such as the neck, chin and jowls. Doubtless, the treatment does have a pleasant effect while being applied, but only the gullible will really believe it's going to reduce the effects of age.

*'Thinking outside the botox.'*

Left: *Metal forks are struck on a solid base to achieve the vibrating effect, and then applied to different parts of the body. The resonance is supposed to release blocked energy and make taut muscles relax.*

The effect of the forks is to create a pleasantly tingling vibration. The therapist usually experiences a pleasantly tingly feeling too, mentally calculating another fat fee on its way.

Many therapists use seemingly hi-tech machinery to create the vibrations. Others take a more direct approach and just hit a mallet over the end of the fork.

# *BUOYANT SWIMSUIT*

These days, buoyancy swimsuits are a matter of controversy at a competitive level. Years ago, however, swimsuit technology was being deployed with the more mundane intention of simply helping people learn to swim, or just to lark about in the water. There was no end of things they tried to come up with – inflatable trucks, inflatable vests, sometimes entire floating suits. Some costumes were quite stylish, others considerably less so. But none of them caught the public imagination. Can you guess why?

*'Holiday snaps are never boring when you're wearing a buoyant swimsuit.'*

Left: *Scenarios such as this, of course, assumed that a lady walking poolside in a dress could always be counted upon to be wearing a swimsuit beneath.*

You could usually guarantee that the model for a buoyant swimsuit would be a shapely female.

She wasn't laughing two minutes later when she was thrown kicking and screaming head first into the deep end.

91

# CALM MACHINE

Scientists have known for years that the electronic frequencies of the brain run in rhythms akin to those of flashing lights, hence the arrival of the mind machines. They generally amount to little more than a pair of goggles hooked up to a 'calm box' which, through a mix of light and sound, are claimed to be able to lull you into a relaxed state.

While there's more than a grain of truth in the scientific thinking upon which these devices are founded, they are rather like those little books claiming to teach you to speak Spanish in a week. Zen-like states of calm are indeed achievable even in our frantic, time-poor 21st-century world, but they are not mastered overnight, nor are they found by plugging yourself into a machine.

*'I tried one and thought I was having an epileptic episode.'*

Left: *You've got to be an optimist if you think donning a pair of dark glasses and switching on a machine is going to introduce new states of calm into your life.*

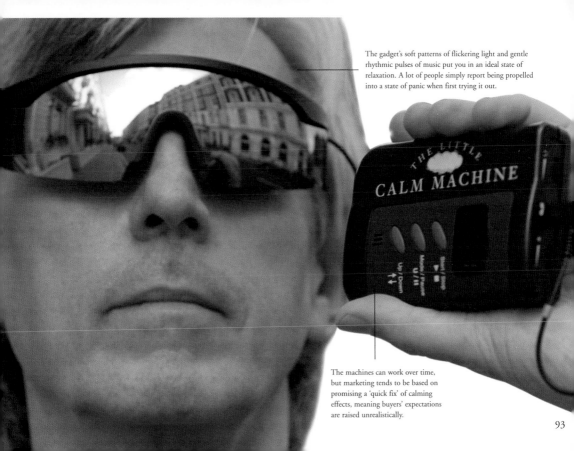

The gadget's soft patterns of flickering light and gentle rhythmic pulses of music put you in an ideal state of relaxation. A lot of people simply report being propelled into a state of panic when first trying it out.

THE LITTLE
CALM MACHINE

Start / Stop
Mode / Pause
▶/II
Up / Down
↑↓

The machines can work over time, but marketing tends to be based on promising a 'quick fix' of calming effects, meaning buyers' expectations are raised unrealistically.

93

# ELECTRONIC CIGARETTE

The 21st century is a hostile place for smokers, forbidden from lighting up just about anywhere except in their own bedrooms. So the electronic cigarette which provides a nicotine hit without producing smoke, adeptly sidestepping the ban, sounds like a good idea. The tobacco-free e-cigarette looks like a conventional one, but contains a rechargeable battery, and when it is inhaled, heats up liquid nicotine and turns it into smoke-like fumes. This means you can sit there nostalgically puffing away in a public place in the sophisticated smoker's pose of yesteryear. The trouble is, these things taste more like dirt than ordinary cigarettes, and in any case, you are still likely to be told off by restaurant staff, public transport officials or members of the public who are unable to tell the difference. Lastly, one of the joys of smoking is the simple act of pulling a cigarette out of the box and lighting up. Sucking on something that you've just charged up like an iPod is unlikely to feel like much of a substitute.

*'Claims to offer a poison-free puff, but still tastes like dirt.'*

Left: *The electronic cigarette offers smokers the prospect of getting their nicotine fix in public without falling foul of the smoking ban. But the experience is unlikely to be peaceful or pleasant, given most people will think they're puffing away on the real thing.*

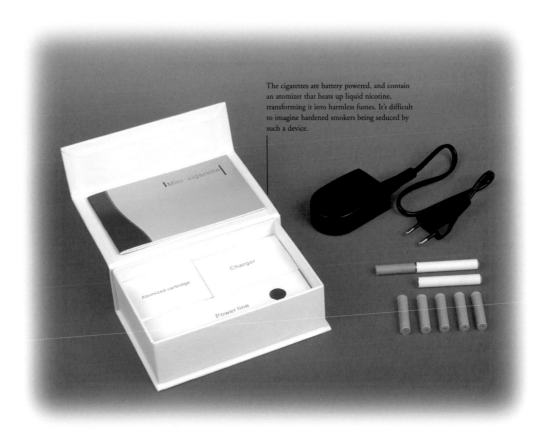

The cigarettes are battery powered, and contain an atomizer that heats up liquid nicotine, transforming it into harmless fumes. It's difficult to imagine hardened smokers being seduced by such a device.

Mini - cigarette

Charger

Atomized cartridge

Power line

95

# ELECTRIC READING GLASSES

E arly in the last century everything suddenly seemed to go electric. Electric lighting replaced gas in homes, the streetcars were powered by electricity, and even your local fleapit cinema was often called The Electric. You might have thought that using electricity for reading glasses was going a bit far until you realized that it was being used to power little lights on the lenses, so the wearer could read in the dark. What an excellent idea this was for the person with irritating relatives who were forever dropping by for a

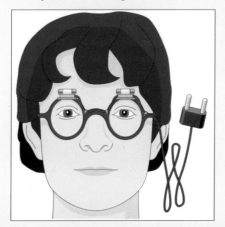

visit. Say you fancied an evening alone. You could switch all the lights off, so that the place looked deserted, while you settled in for a night of uninterrupted reading. Sheer bliss!

*'Best worn when home alone.'*

Left: *The glasses enjoyed a short-lived comeback during the energy crisis of the 1970s.*

You could sit wearing the glasses and pretend that you were at the opticians.

An upright posture had to maintained at all times, and it was most annoying when the meter ran out.

97

# *ENERGY COCOON*

Qi (pronounced 'Chi') is an invisible life force that, according to
Chinese philosophy, helps you stay alert mentally and physically. If it
becomes blocked or inhibited, however, you may become unwell. But you
need never worry about that happening if you are rich enough to splash out
on an Energy Cocoon, which operates like an all-in-one mini sauna. Its
multiple lamps (orange, yellow, green, then back to orange) wrap you in
soothing 'chromotherapeutic' rays, while Jacuzzi-like water jets massage your
entire body. It finishes off by giving you a shower. What's not to like? It looks
a bit lightweight, though, so be careful you don't nod off, accidentally tip it
over, and flood out that nice old couple in the apartment below.

**RATINGS**

| | |
|---|---|
| **STYLE COUNT:** | ★★★☆☆ It's a curvaceous little number, but how many people would have a bathroom big enough to accommodate it? |
| **ORIGINALITY:** | ★★★☆☆ |
| **USEFULNESS:** | ★★★☆☆ |
| **NERD APPEAL:** | ★☆☆☆☆ |
| **LONGEVITY:** | Could go down well with flashy stockbroker types. |

*'Looks like a giant toilet.'*

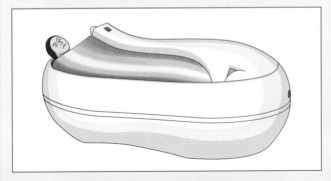

Left: *It always helps if you are young, lithe and flexible when getting into these things.*

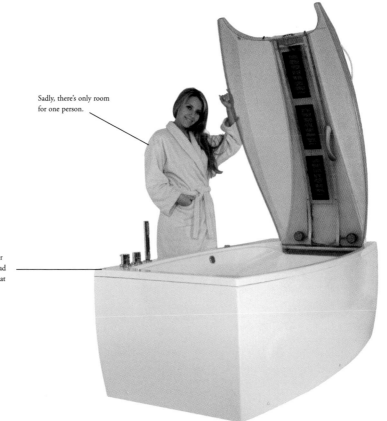

Sadly, there's only room for one person.

As long as you are under 6 feet tall, only your head will be left sticking out at the end.

99

# EYE-DROPPER EYE GLASSES

Using an eyedropper to squeeze medication onto your eyeball is a tender, awkward business at the best of times, but what happens if you are so short-sighted, you can't see what you are doing without your glasses? At such moments, a pair of custom-made eye-dropper eye glasses might seem a good idea. But if you thought they looked basic – just a pair of spectacles with two

small funnels fitted through holes in the lenses to dispense the lubricant – think again. Fitting you up with a pair of the glasses involves taking very specific measurements of the size of your head, from the middle of your nose to the middle of your eye, because everyone's facial anatomy is unique. Then, with the glasses on, it's vital to tilt the head at the right angle to stop the liquid from running off down your cheeks. Ultimately, you'll find yourself falling back on what everyone else does – short-sighted or not – just taking aim, squeezing the dropper and hoping for the best.

*'Do I look odd in these?'*

Left: *Keeping your eyes open long enough for the liquid to reach them is clearly going to be a challenge using a pair of eye-dropper glasses.*

100

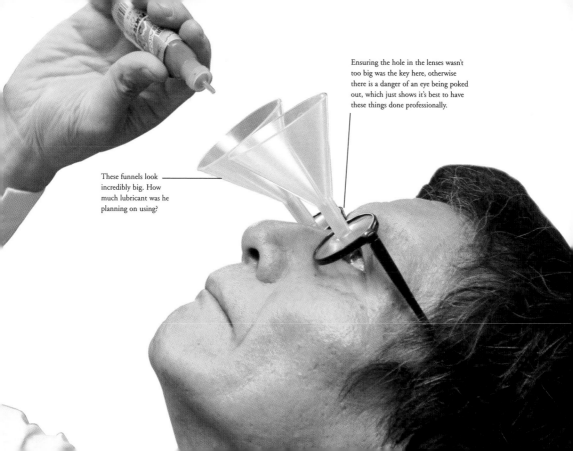

These funnels look incredibly big. How much lubricant was he planning on using?

Ensuring the hole in the lenses wasn't too big was the key here, otherwise there is a danger of an eye being poked out, which just shows it's best to have these things done professionally.

# FIREPROOF SMOKING DEVICE

In the 1950s, the link between smoking and cancer was still not fully backed up by scientific evidence. The main fears about the activity revolved around the possibility of injury to people or property caused by smoking-related fires. One preventative measure that was devised was a special ashtray that you could attach to the length of the cigarette to catch the ash. But the most ridiculous measure anyone ever came up with was that of smoking via a tube, with the cigarette attached to an opening at the end, and resting in an ashtray. While the idea was to enable you to fall asleep without risk of causing a fire, the inventor seemed to overlook the more obvious danger of choking to death.

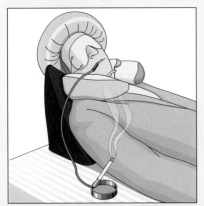

**RATINGS**

STYLE COUNT: ★★★★★
ORIGINALITY: ★★★☆☆
USEFULNESS: ★★★★★
NERD APPEAL: ★★★★★
LONGEVITY: A stupidly dangerous idea still aimed at people with health problems.

*'Stop killing yourself!'*

**Left:** *The inventor of the fireproof smoking tube had laudable intentions, but seems to have introduced a new health risk from smoking: death by choking.*

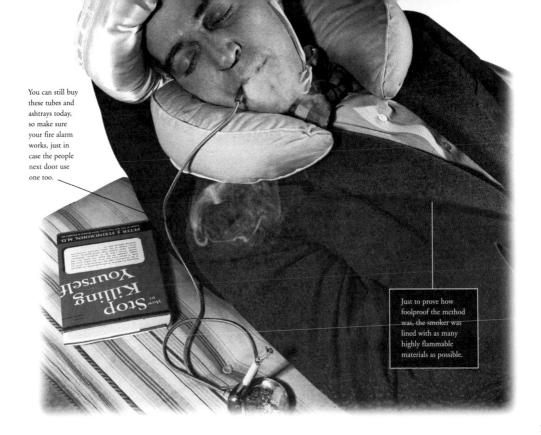

You can still buy these tubes and ashtrays today, so make sure your fire alarm works, just in case the people next door use one too.

Just to prove how foolproof the method was, the smoker was lined with as many highly flammable materials as possible.

103

# *FLOTATION SUIT*

Flotation swimsuits were a gimmick designed to cater for the family seaside holiday in the days when, it seems, while there may have been less flab on view than in the modern era, people were still oblivious to how absurd they could look in unsuitable garb. The most popular types were ones with shoulder pieces, ideal for family fun, splashing around in shallow water. With just a few hearty puffs into the pneumatic pads, you could paddle away to your heart's delight. You had to mind the rocks though, because if you got a puncture, you'd sink like a stone.

**RATINGS**

**STYLE COUNT:** ★★★★★
**ORIGINALITY:** ★★★★★
**USEFULNESS:** ★★★★★
**NERD APPEAL:** ★★★★★
**LONGEVITY:** Still a hit with families today.

*'Pretend you're in the Dead Sea.'*

**Left:** *Inflatable swimwear allows holidaymakers to keep up with those all-important news headlines from back home.*

Before he left for the planet Mongo, Ming the Merciless was just a happy family sort of guy.

The dog, however, wasn't so keen on him being around.

105

# MARINE LIFE-SAVING APPARATUS

The inventor of the marine life-saving apparatus didn't intend just to keep you afloat after a shipwreck. He thought it was a better idea to provide you with the means to swim away. He certainly prepared you for all contingencies. The 'apparatus' was in the shape of a whale or shark, which would obviously fool any potential predators who might otherwise mistake you for a human. There was an electric motor and a propeller, and a valve controlling the admission of compressed air. Just to round things off, the whole thing was collapsible, and could be slung over your shoulder and carried away once you reached dry land.

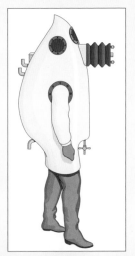

## RATINGS

| | |
|---|---|
| STYLE COUNT: | ★★★★★ |
| ORIGINALITY: | ★★★☆☆ |
| USEFULNESS: | ★★★★★ |
| NERD APPEAL: | ★★★☆☆ |
| LONGEVITY: | Another invention you hope never left dry land. |

*'A suit that's just a drop in the ocean.'*

Left: *The apparatus could be collapsed and carried over your shoulder, but it was probably less bother just to carry on wearing it.*

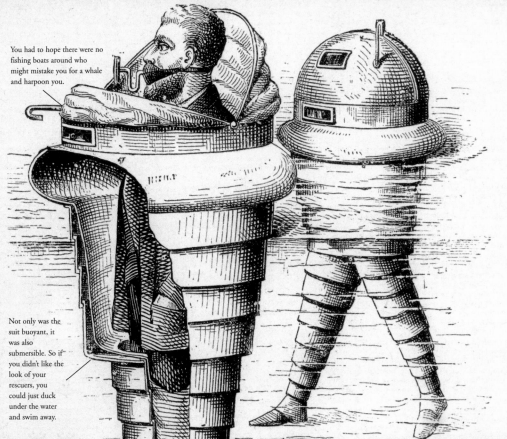

You had to hope there were no fishing boats around who might mistake you for a whale and harpoon you.

Not only was the suit buoyant, it was also submersible. So if you didn't like the look of your rescuers, you could just duck under the water and swim away.

107

# RAILROAD SNOWPLOW

Some of the earliest snowplows were pioneered in the first decades of the last century on the American railroads when, before the development of a decent highway network, travel by train played a crucial role in the young country's economy.

The simplest railroad snowplows worked by pushing the snow to either side of the track, but for heavier falls a rotary snowplow with a huge set of circular blades was used. The snow was thrown up onto the blades, and then through a chute out the back onto either side of the line.

The trouble was that they were expensive to maintain and prone to

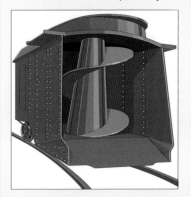

jamming. Private railroad operators were always looking to save money, and these machines, sometimes used only a few days in the year, provided an opportunity to cut corners. These days, rotary plows are museum pieces, but modern plows are just as prone to break down, and you suspect short-sighted stingy economizing remains the root cause.

*'A new-fangled twist on the old-fashioned cowcatcher.'*

Left: *You really would not want to get caught up in the lethal blades of a vintage rotary snowplow.*

Snowfall can bring rail networks to a halt for days on end, so it's strange they often gamble on saving money by being cheap with snowplows.

The snow was blown out onto the sides of the track, but if the wind was blowing the wrong way, it just ended back where it was.

109

# SLIMMING BELT

To look at the advertising for most slimming belts, you'd think all you have to do is fasten one around your waist or hips, and then sink back in an armchair with the TV remote control while the belt does the hard work of ridding you of your excess pounds. The belts operate by pushing currents through the body to raise the temperature, so that the wearer sheds weight. But weight loss through dehydration only has a temporary effect. As soon as you rehydrate by drinking water, your hips and waist expand again. Worse, these belts heat up to high levels, sometimes as much as 42°C (107°F). You can spend much of the time in considerable discomfort, or even suffer burns.

**RATINGS**

| | |
|---|---|
| **STYLE COUNT:** | ★★★★★ |
| **ORIGINALITY:** | ★★☆☆☆ |
| **USEFULNESS:** | ★★★★★ |
| **NERD APPEAL:** | ★★★★★ |
| **LONGEVITY:** | As long as people think there's an effort-free way of losing weight. |

*'Gives new meaning to the term "burning calories".'*

Left: *They're not called 'melt belts' for nothing. They can rise to such high temperatures that users have been burned.*

Currents are passed through the body to create heat or by raising the wearer's own heat levels. But apart from the fact that they don't really work, many users admit they cause more discomfort than a session in the gym with a sadistic personal trainer.

The belts are said to help to shed weight around the abdominals, back, waist, hips and even shoulders.

# *SMART GOGGLES*

The inventor of smart goggles aimed to banish 'senior moments', the euphemistic term for forgetting where you last put items like your reading glasses, wallet or car keys. The goggles contain a built-in camera, a screen and a small computer. The camera records everything it sees, and stores the information on the computer's sophisticated object-recognition software. If you lose an item, you simply 'ask' the computer where it last recorded seeing it, or play back the film on the screen. The defects here are obvious. The built-in web cam makes the goggles bulky and uncomfortable to wear and, if they are to be of any use, you need to wear them constantly. But they won't do you much good if another member of the household has moved the lost object. Taking all this into consideration, smart goggles are probably just too 'smart' for their own good.

*'Great! I found my keys. Now all I need to worry about is looking like a total idiot.'*

Left: *The camera is fitted on the right lens of the glasses and records most things in front of it. And just to make wearing the goggles even more uncomfortable, a web cam is mounted above the left frame.*

112

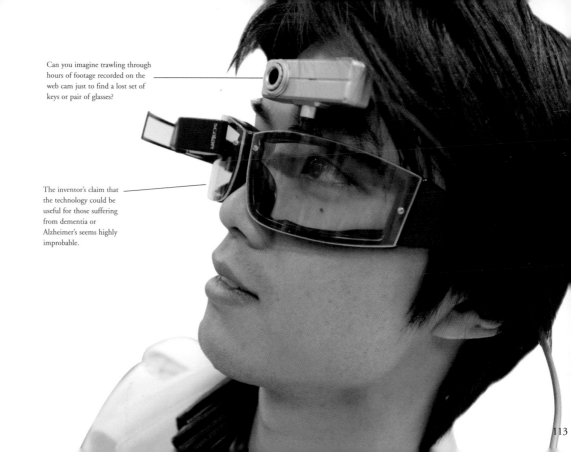

Can you imagine trawling through hours of footage recorded on the web cam just to find a lost set of keys or pair of glasses?

The inventor's claim that the technology could be useful for those suffering from dementia or Alzheimer's seems highly improbable.

# EVERY HOME SHOULD HAVE ONE

The pleasure of possessing a home full of modern conveniences was a phenomenon born in the early 20th century. Stimulated by a burgeoning advertising industry, inventors and technicians found a new market to exploit. While this has led to some ingenious time-saving devices, others have been no more than desperate attempts to cash in. The more dubious inventions would have you donning a pair of brightfeet slippers when you arose in the night guiding you along the hallway to a toilet lit by a nightlight. When morning came, you'd be roused by a 'silent' alarm-clock, and the bed would make itself. You'd smile smugly as your guests remarked on your cheerful temperament, brought on by the wisecracks printed on your toilet paper and the fact that you were the proud owner of an automated lawnmower that cut your grass for you.

In reality, none of these items are of any real use around the house – but helping you was actually never on the agenda for these inventors. The sole aim of just about every invention in this section (except perhaps the much mocked 8-track tape), was simply about getting you to part with your hard-earned cash.

Left: *The lazy occupant of this deckchair clearly wasn't planning on going anywhere soon. But then, probably neither was the remote-controlled lawnmower.*

# 8-TRACK TAPE CARTRIDGE

From the mid 1960s to the early 70s, 8-track tape cartridges seemed a breakthrough invention, primarily for playing recorded music while driving. Even in homes it seemed set to challenge the dominance of vinyl records. All you had to do was push a tape into the player and away it went on a continuously playing loop. But the 8-track was flawed. If you wanted to hear a song again you couldn't rewind the tape, but had to wait until it looped around once more. Tapes wore out quickly and jammed easily. When the revolving tape heads moved out of alignment, it caused audio bleed, which made it sound like two songs were playing at once. As soon as the smaller, more reliable audio cassettes came along in the mid 70s, 8-track was on its way out.

*'I didn't think 8-track meant hearing 8 songs all at once.'*

**Left:** *In the 1960s, being able to choose what music you listened to on the road was quite a novelty. The reliability issues associated with 8-track, however, probably had most drivers sticking to the radio.*

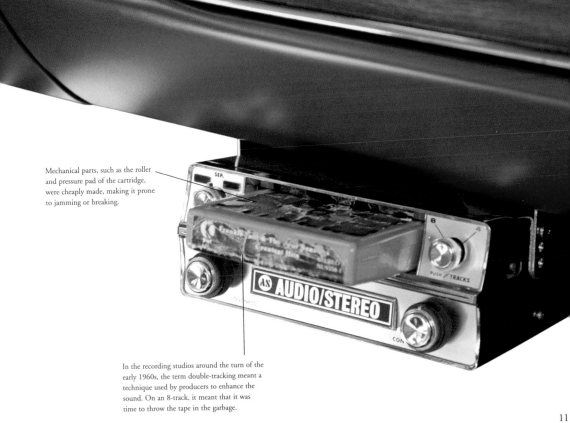

Mechanical parts, such as the roller and pressure pad of the cartridge, were cheaply made, making it prone to jamming or breaking.

In the recording studios around the turn of the early 1960s, the term double-tracking meant a technique used by producers to enhance the sound. On an 8-track, it meant that it was time to throw the tape in the garbage.

# AQUATIC POD SUITE

While it may look as if it's a sign that the aliens have arrived, the Aquatic Pod Suite is really just an exceptionally grandiose, static, houseboat. Its flying saucer shape disguises a 14m sq (150ft sq) interior full of the conveniences you'd expect in a plush hotel. Surrounding it is an inflatable floating terrace. As well as aiding buoyancy, the terrace provides a viewing platform, though you are as likely to spend as much time inside, where beams directed at the see-through bottom allow you to watch the aquatic world beneath your feet. Just in case someone in a rusty old boat berths alongside you, the inflatable terrace provides protection from bumps or scratches. But since only a would-be villain in a James Bond movie would own something like this, you'll probably be intending to feed anyone who comes too close to the sharks anyway.

## RATINGS

STYLE COUNT: ★★☆☆☆ Looks a little too garish

ORIGINALITY: ★★★☆☆

USEFULNESS: ★★☆☆☆

NERD APPEAL: ★★☆☆☆

LONGEVITY: Shouts 'money' a little too obviously to have much appeal.

*'A Christmas stocking stuffer for the man who has everything.'*

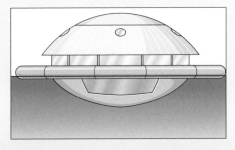

**Left:** *You'll get some funny looks when the Aquatic Pod Suite is towed into the docks. But if you own one of them, you're probably far too wealthy to care.*

You can sit back and enjoy a full 360 degree view.

There's also a see-through bottom, making it perfect for mermaid spotting.

# BETAMAX

In the mid 1970s, Sony, market leaders in consumer electronics, seemed set to dominate on a new front – home video recording systems that would revolutionize the way people watched television. Unfortunately, the company backed the wrong horse by choosing the Betamax format, and was quickly outflanked by rival manufacturers who opted for the more versatile VHS. The main problem was that Betamax tapes were limited to a maximum of sixty minutes of recording time. This may have been the average length of an American TV show, but it was inadequate for feature films. When VHS tapes four times the length arrived, there was only one choice as far as the public was concerned, even if the picture quality and tape durability were inferior. Additionally, the machines that played Betamax tapes were very pricey and beyond the financial reach of a large part of the target market. By 1988, Sony had hoisted the white flag and began producing VHS format recorders.

**RATINGS**

STYLE COUNT: ★★★☆☆

ORIGINALITY: ★★★☆☆

USEFULNESS: ★★★☆☆

NERD APPEAL: ★★★★★

LONGEVITY: A losing cause almost from the moment it hit the stores.

*'We're better than normal people. We've got a Betamax.'*

*Left: The idea was to produce a tape the same size as a Sony company diary, but the Betamax proved to be the Beta of ideas when the VHS became market leader.*

You can still buy Betamax tapes because diehards argue that, for quality, they continue to surpass anything subsequently available on the market.

# BOOSTER BLADES

Booster blades are kind of a cross between a bicycle and a conventional pair of in-line roller skates. They are worn like normal skates but you propel yourself along in a stepping or pedalling motion. The wheels operate via gears, sprockets and a chain, similar to bike technology. Braking, however, is done the same way as on a skate by simply tilting either foot backward onto the ground. The inventor claims that the chain mechanism gives you greater propulsion than on a pair of normal skates. One thing is certain – you'll look more fool than cool clopping along on these clumsy objects.

### RATINGS

**STYLE COUNT:** ★★★★★ Are these not the ugliest things on the planet?

**ORIGINALITY:** ★★★★★

**USEFULNESS:** ★★★★★

**NERD APPEAL:** ★★★★★

**LONGEVITY:** Unlikely to pose a threat to more established methods of blading.

*'How to kill yourself in thirty seconds.'*

Left: *She looks like she could walk faster. But once that old adrenalin starts surging, how long before she leans forward into a curve, and turns herself into a skating deathtrap?*

Movement is based on simulating the pedalling stroke of cycling, with the boot plate tipping forward as the foot is lifted. Of course, this looks quite odd, given the absence of a saddle to sit on.

When an idea gets this fussy, you know that it's dead in the water.

123

# BRIGHTFEET SLIPPERS

The inventor of the brightfeet slippers says his idea came in a moment of inspiration. There he was in bed one night. He woke up, threw back the covers and, in fumbling around for his slippers, stubbed his toe. It was at that second that he realized that there must be millions of others just like him, who'd got up and done the same thing. What they all needed, he reasoned, was a slipper that lit up.

So now we have the brightfeet, with little LEDs that only come on when you slip your feet inside. For added safety they are nonskid, so you can wander around your dark old house in the middle of the night as much as you like. Of course, given they shine a light fully 8m (25ft) in front of you, your nocturnal ramblings are likely to wake everyone else in your household too. But who cares, now you need no longer worry about stubbing your toe?

### RATINGS

STYLE COUNT: ★★★★★

ORIGINALITY: ★★★★★ Yes, we'll give them that.

USEFULNESS: ★☆☆☆☆

NERD APPEAL: ★★★★★

LONGEVITY: Can imagine these going down well – and provoking raised levels of profanity – in an old folks' home.

*'You'll never lose yourself in the dark again.'*

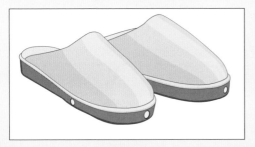

Left: *Powerful bright lights only come on when you put your feet inside. The trouble is, it's while trying to find the slippers that toe stubbing incidents are most likely to occur.*

124

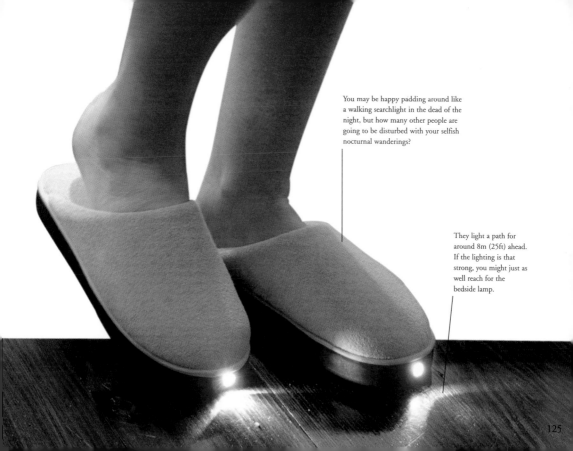

You may be happy padding around like a walking searchlight in the dead of the night, but how many other people are going to be disturbed with your selfish nocturnal wanderings?

They light a path for around 8m (25ft) ahead. If the lighting is that strong, you might just as well reach for the bedside lamp.

125

# CAMERA GUN

Though they may have looked like they could blast a hole through the steel door of a safe, gun cameras in their various guises were actually perfectly harmless objects, even if they were deployed in World War I and II. Mounted on military aircraft, the idea of integrating the camera with a barrel was supposed to ensure a steady 'shot', as it did in rifle technology. It also took pictures on the plane that would be used to assess the standard of gunnery in the air. Today they are collectors' items, but whether anyone has ever had the nerve to stand still long enough while their picture was taken by a camera gun is unknown.

**RATINGS:**
STYLE COUNT: ★★★☆☆
ORIGINALITY: ★★★★★
USEFULNESS: ★★★☆☆
NERD APPEAL: ★★★★☆
LONGEVITY: Have been used in at least three major conflicts.

*'Don't shoot me – I'm just the photographer!'*

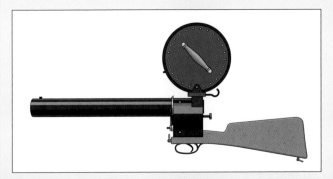

Left: *Even Leica used its considerable expertise to design a 400mm camera gun.*

More recent gun cameras were used by military reporters in the Vietnam War, though it seems a strange idea to enter a war zone as a noncombatant possessing something that looked like an offensive weapon.

The first guns used during World War I took pictures using 120mm roll film and were styled to match the design of the Lewis Gun. The pictures would be used to determine the angle and range of gunshot.

# CHESS FOR THREE OR FOUR PLAYERS

There are almost 1500 known variants on the standard game of chess, but none has ever succeeded in knocking the original format off its premier pole. There have been experiments with different pieces and moves and, most recently, innovations that took the heretical step of placing the back row pieces on the wrong squares.

The most popular variant is chess for three or four players. Three players, however, are generally inadvisable – it often means two of the participants gang up to concentrate on attacking the third player, followed by a bitter, undignified brawl between this pair to be the first to actually call 'checkmate' and be declared the victor.

Four-handed chess is little better, because everyone starts attacking everyone else in an almighty free-for-all, and pieces start flying all over the place. In the middle of this bedlam, it's not unusual for two of the players to quietly slip away from the table, and set up their own game of – yes, you've guessed it – good old ordinary chess.

*'I've just been stabbed by the knight.'*

Left: *Between two sides, chess is a classic game of tactics and conflict. Played among three sides and it descends into a stalemate of arguing over the rules, name-calling and sulking.*

128

When it comes to chess, two's company, three's a quagmire.

These wild and crazy people are either trying to play four-handed chess with just three players, or things turned nasty and the fourth player is en route to the hospital.

129

# CLACKERS

All you have to do to ensure your invention goes down in history is to make something that achieves a short period of fame, and then suddenly gets banned. Such was the fate of clackers. For a brief period in the early 70s, these were the kings of the toy stores, yet all they consisted of was a pair of hard plastic balls or marbles, attached to a sturdy piece of string on a ring. You put your finger through the ring and tried to get the balls to bang – or clack – against each other. Oh, the happy hours that were spent in the playground trying to master these things. But then killjoys decided kids could get hurt because if they clacked the balls too hard, the marble could shatter, inflicting unpleasant eye injuries. Now they're just a nostalgic exhibit in a museum.

**RATINGS**

STYLE COUNT: ★★★★★

ORIGINALITY: ★★★★★

USEFULNESS: ★★★★★

NERD APPEAL: ★★★★★

LONGEVITY: Just a noisy, but fun couple of years.

*'The child's toy that doubled up as a lethal weapon.'*

Left: *Clackers were popular with children and teenagers. Adults complained about the tediously repetitive clacking sound, and so had more than one reason to delight in the ban.*

Of course, if the string broke, the balls had an alternate use as missiles to be lobbed at someone you didn't like.

Masters of clacker art could get the balls to hit each other so hard they'd reverberate in a full, spectacular circle.

131

# THE :CUECAT

Almost from the moment it became available in 2000, the :CueCat was being written off as, at worst, a waste of time and, at best, an interesting toy. Designed in the shape of a cat (complementing the computer mouse) it was marketed as a cheap but effective barcode scanner that would act as a bridging tool between printed and online advertising media. Users could attach the wired :CueCats to their USB ports to read barcodes found in magazines, newspaper ads or on business cards, which would connect them to related websites.

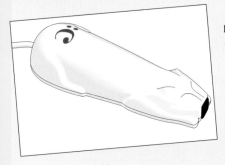

But, it was quickly pointed out, few home computer users were going to take the trouble of installing the scanners. Then news of a security breach that gave away private information about users, including their names, email addresses and zip codes, sounded the death knell for the :CueCat. By 2005, the makers of the product, the Digital Convergence Corporation, had gone into liquidation.

*'It fails to solve a problem that never existed.'*

Left: *The :CueCat arrived on the scene around the time of the dotcom boom. Its success was just as fleeting.*

The :CueCat was shaped like a cat, but acted like a mouse.

The idea was that when the :CueCat was wired up to a computer USB port, it could read magazine or product barcodes, which would connect users to related websites.

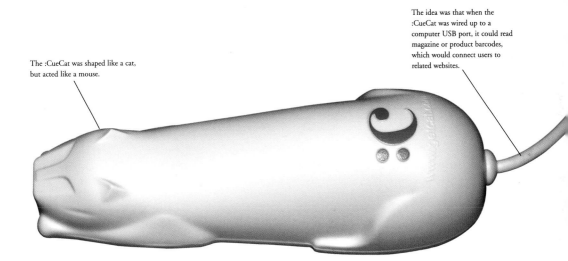

133

# DOGGIE SWEATER

No one remembers the inventor of the doggie sweater, but whoever it was knew what they were doing because they found a market that keeps on growing. There's no doubt that doggie apparel (a Boston terrier in a nifty leather jacket, or a chihuahua in a cute little T-shirt) is good for a laugh. But isn't it a little cruel and unnecessary? Actually, there could be good reasons for dressing up your pet in a doggie sweater. If it has just undergone treatment that caused hair loss, the vet may have advised some temporary protection from the elements. Also, short-haired dogs, in particular, can be prone to feeling the winter chill. Of course, in their natural states all dogs had thick fur coats, but pedigree bloodlines have in many cases bred these characteristics out. Still, you have to wonder whether the latest development of all this, including hats and sunglasses to protect dogs from UV rays, isn't going a bit too far.

*'If you really want to dress something up, go buy a doll.'*

Left: *It can occasionally make sense to add a warm layer of clothing to your dog on the advice of a vet, but most times, it looks like it is just indulging the whims of a sad owner.*

A short-haired dog, thanks to decades of selective breeding, can genuinely feel the cold in the winter.

Where it gets annoying is when you see the poor creature forced to wear the sweater in summer.

# EARLY HANDS-FREE PHONES

Cordless phones may seem like a revolution dating from the 1990s, but the technology for them was making headlines as far back as the 1930s. It was the weight of early receivers that most exercised the minds of inventors of that period, however. Prewar telephone receivers were cumbersome, weighty objects to hold for long periods, so you might think that new extension arms, or 'third hands', would have been well received by office workers. By attaching extensions to the receivers in the form of flexible tubes on lamp-style stands, adjusted to hold the receiver to the ear, the user's hands

were left free. The trouble was that, while they certainly freed up both hands, the head had to be held absolutely still. By the end of the working day, you had simply traded your aching arm for a very stiff neck.

**RATINGS**

STYLE COUNT: ★★★★★ The lampstand affair was a tidy looking setup.

ORIGINALITY: ★★★★★

USEFULNESS: ★★★★★

NERD APPEAL: ★★★★★

LONGEVITY: File under early telecommunications curiosities.

*'Great for perfecting the "academic's stoop" look.'*

Left: Which office worker hasn't wished they had a third hand? But unless you got the position of the tube arm absolutely right, these gadgets were a mixed blessing.

The idea of putting your receiver on 'speakerphone' is clearly nothing new. She doesn't look like she had much fun using it, though, does she?

There must have come a point when you dreaded picking the receiver up and hearing the operator say: "Will you accept a long-distance call?"

# *EASY-BAKE OVEN*

The Easy-Bake Oven, first launched in 1963 to allow little girls an early dip into home-style baking, is still popular today, so there's no denying its appeal over the generations. All the child has to do is follow the instructions on the Easy-Bake cake mixes, and pop the recipe inside. A window allows them to watch the cake baking. Tension mounts, and most likely, parents start to feel queasy. Doubtless there are many who can only smile with pleasure at the mention of the Easy-Bake Oven of their childhood. But for parents, the recollection of the burnt offerings they were obliged to taste, and smile encouragingly as they forced themselves to swallow, must make them thankful for the passing of time.

Easy-Bake

**RATINGS**

STYLE COUNT: ★★★★☆

ORIGINALITY: ★★★★★

USEFULNESS: ★★★★☆

NERD APPEAL: ★★☆☆☆

LONGEVITY: Five decades and still going strong.

*'The first oven was a lurid turquoise, and the cakes didn't look too good either.'*

Left: *The Easy-Bake Oven is a charming idea, and has sold in its millions. Unfortunately, while the kids have a lot of fun baking in them, parents forced to sample the outcome, come to view them as objects of terror.*

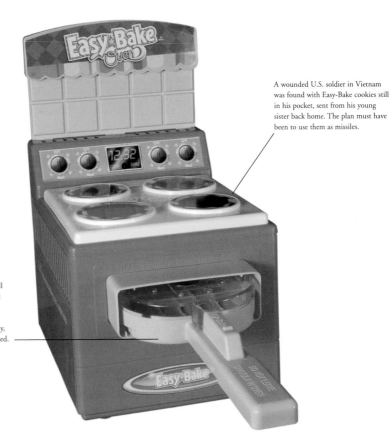

A wounded U.S. soldier in Vietnam was found with Easy-Bake cookies still in his pocket, sent from his young sister back home. The plan must have been to use them as missiles.

The early engineering was supposedly inspired by the small ovens used by New York pretzel vendors. But because the ovens work on heat provided by lightbulbs, rather than electricity, the food tends to be undercooked.

# FLOATING BED

It could be the nearest we've yet come to a magic carpet made real, were it not for the fact that contact is actually maintained with the floor by four thin cables. However, the floating bed truly does seem to hover in mid air. It derives this ability from the power of opposing industrial-strength magnets, above and beneath the bed. It certainly looks like the kind of ultramodern fitting that you might see gracing some swanky uptown apartment. But in the end it just smacks of expensive flimflammery. Who really needs to sleep in a floating bed?

*'Makes life easier for the cleaners.'*

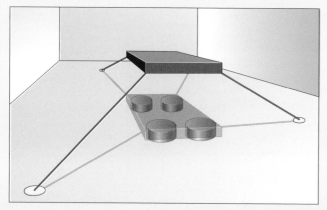

Left: *The bed hovers thanks to opposing industrial-strength magnets. Not great news if you have a pacemaker.*

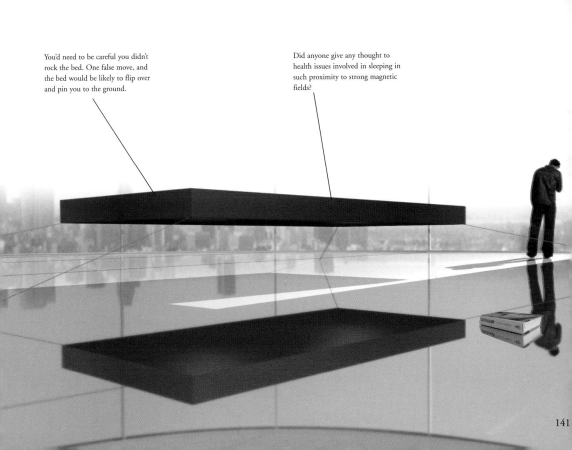

You'd need to be careful you didn't rock the bed. One false move, and the bed would be likely to flip over and pin you to the ground.

Did anyone give any thought to health issues involved in sleeping in such proximity to strong magnetic fields?

141

# *GLOW-IN-THE-DARK SUNGLASSES*

Sunglasses may have been technically designed to protect the eyes from the glare of the sun, but ever since they first became commercially available in the late 1920s, they've had an irresistible allure for the fashion-conscious. Glow-in-the-dark sunglasses are simply the latest, tackiest manifestation of this. Completely useless, providing no protection at all from UV rays, they are a nightwear accessory, appealing mainly to young clubbers and partygoers. Batteries illuminate the frames in brightly glowing shades from

pink to green, either in static or flashing mode. More popular, however, are nonbattery powered shades that simply emit a soft glow when exposed to back or strobe lighting.

**RATINGS:**

STYLE COUNT: ★★★★☆
ORIGINALITY: ★★☆☆☆
USEFULNESS: ☆☆☆☆☆
NERD APPEAL: ☆☆☆☆☆
LONGEVITY: A fleeting appeal for fashion victims.

*'The shades that don't work in the sun.'*

Left: *Most sunglasses, however low the filter range, provide at least some eye protection. Glow-in-the-darks dispense with any such pretence and are worn solely for effect.*

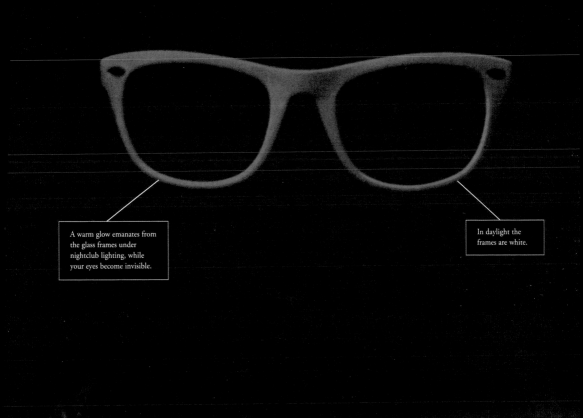

A warm glow emanates from the glass frames under nightclub lighting, while your eyes become invisible.

In daylight the frames are white.

# *HULA HOOP*

The humble Hula Hoop is so simple it seems to date back to time immemorial. Egyptian children fooled around with large hoops made from dried grapevines, and the ancient Greeks used 'hooping' as a form of exercise. Children have made them out of wood, grasses and bamboo for

thousands of years. The actual name Hula Hoop only seems to have come about after British soldiers, landing on Hawaii in the early 1800s, saw the resemblance of the native hula dancing to hooping.

However, the Hula Hoop craze really took off in the late 1950s when a company started making them out of plastic. Popularity spread so fast that over a million were sold in the USA in one summer alone in 1958. The sales made millions for the manufacturing company Wham-O, thanks to a remarkably slick piece of marketing, considering the objects had been around for centuries.

## RATINGS

STYLE COUNT: ★★★☆☆

ORIGINALITY: ☆☆☆☆☆

USEFULNESS: ★★★★★ A cheap way of staying in shape.

NERD APPEAL: ★☆☆☆☆

LONGEVITY: Nearly as old as the oceans.

*'It's a lot cheaper than a Wii™.'*

Left: *While Wham-O made millions selling plastic Hula Hoops in the late 1950s, the inventors made nothing because you couldn't patent an idea that had been around for centuries. More recently, hooping has regained its popularity, and Hula Hoop competitions are held around the world.*

Hula Hoops are a great way of a shedding a few pounds around waist and hips. Shame they are wasted on the young.

The suggestive hip action led to Indonesia banning them as unseemly.

147

# LOUNGE HOVER CHAIR

The designer of the lounge hover chair claimed it was inspired by the *Star Wars* movies. It certainly comes across as a spaced-out idea. Made by hand and not machine, it's said to defy gravity with the use of repelling magnetic forces built into the design, so that as you lie back and close your eyes, you feel as if you are floating on a cloud. Suddenly, though, you jolt bolt upright again with the horrific realization that you've just spent an extortionate amount of money on an item that is hideous as a dentist's chair and of considerably less usefulness.

**RATINGS**

STYLE COUNT: ★★★☆☆
ORIGINALITY: ★★★☆☆
USEFULNESS: ★☆☆☆☆
NERD APPEAL: ★★☆☆☆
LONGEVITY: You can imagine it appearing at a museum exhibition on early 21st century interior design in a 300 years' time.

### 'Seat belts not provided.'

Left: *As a once-only experiment the hover chair is certainly worth a look, and the idea of supplying loungers in clear acrylic, so that you can see its components, is commendably honest. But that doesn't make it any prettier.*

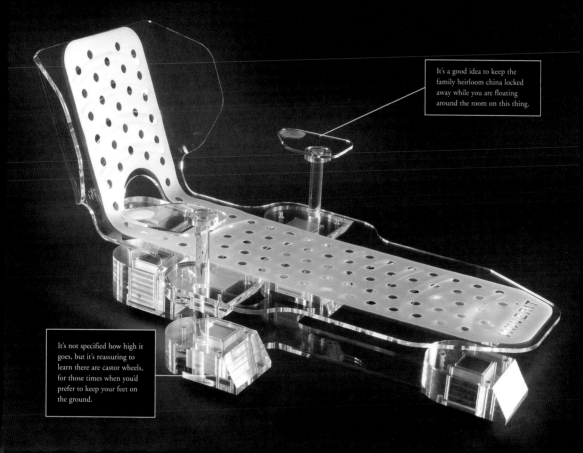

It's a good idea to keep the family heirloom china locked away while you are floating around the room on this thing.

It's not specified how high it goes, but it's reassuring to learn there are castor wheels, for those times when you'd prefer to keep your feet on the ground.

# MOTORIZED ROLLER SKATE

Motorized roller skates have a surprisingly long pedigree. As far back as 1906, Henry Beauford was touring the western U.S. states with his 'motor powered skate'. Once the motor started, apparently, nothing other than the skater's movement was required to achieve propulsion. Then in 1956, Antonio Pirrello invented his gas-powered skate, which operated via a gasoline motor that was worn as a backpack. The skates were said to roll away at up to 65km/h (40mph), so it's hardly surprising that, despite a fair bit of publicity in lifestyle magazines of the time, Mr Pirrello's invention soon faded into obscurity.

**RATINGS**

STYLE COUNT: ★★★★★
ORIGINALITY: ★★☆☆☆
USEFULNESS: ★★★★★
NERD APPEAL: ★★☆☆☆
LONGEVITY: Motorized skates have been banned virtually worldwide, but black market availability means you could probably still get your hands on a pair.

*'Roller skating by jet power.'*

Left: *The gas-powered skate could reach speeds of 65km/h (40mph), just the thing for taking baby out for a quick spin.*

The gas connected to the right skate, with steering achieved through the left foot. A second cable was connected to a hand-held clutch, thus regulating the speed.

More recent motorized skates of the 21st century, which actually have small fuel tanks attached to the skater's leg, have been banned on safety grounds.

149

# PET LOO

Up to a point it's hard to argue with the logic of an indoor pet comfort station, tailor-made for the needs of the dog that lives in an apartment building. Indoors all day, while the owner is out at work, or spending an evening locked up at home while the owner heads off for a night on the town, the dog, as soon as it feels the need to do a number one or a number two (and having earlier mastered the necessary training formalities), just heads for this stretch of make-believe interior turf. On coming home, the owner need only flush the waste away, wash the tray down with hot water and everyone is happy. The owner has probably wasted a lot of money on something they could have devised themselves, but if you're a dog whose misfortune is to be kept indoors in this way, you deserve every perk available.

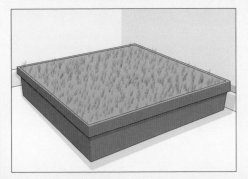

## RATINGS

STYLE COUNT: ★★★★★
ORIGINALITY: ★★★★★
USEFULNESS: ★★★★★
NERD APPEAL: ★★★★★
LONGEVITY: A sure winner, with half the world's population now living in a city high-rise (and keeping pets) this invention is here to stay.

*'Stench-free (at least for the dog).'*

Left: *Dogs are easily trained to use the pet loo. But you suspect it could quickly become the easy option for the lazy owner who can't be bothered to take their dog for a walk.*

This dog looks a bit lonely, but how would you feel with just a pet loo for company all day?

Providing the dog with a bit of artificial lawn to go on is a thoughtful touch, considering it's probably the nearest they're going to get to a glimpse of green grass.

# PET UMBRELLA

As anyone who has kept a dog with a dislike of water will agree, the idea of a pet umbrella isn't quite as derisive as it sounds. Many a normally boisterous, outdoor-loving dog will gloomily take to its bed rather than face a walk in the rain. The pet umbrella is made of waterproof fabric and, with

an automatic pop-up button, is simply attached to the dog's collar. It's foldable, and is used with a normal leash. There are two snags here. Many dogs are going to be frightened by having this shadowy thing hanging above them and will try to shake it off. Meanwhile, the umbrella is only designed for small dogs. So Fido the Labrador will probably end up staying home, thankfully curled up asleep, after all.

## RATINGS

| | |
|---|---|
| STYLE COUNT: | ★★★★★ |
| ORIGINALITY: | ★★★★★ |
| USEFULNESS: | ★★★★★ |
| NERD APPEAL: | ★★★★★ |
| LONGEVITY: | Likely to have appeal restricted to pampered uptown pooches only. |

*'Comes with owner umbrella in matching pattern.'*

Left: Long-suffering cart horses on hot city streets may appreciate it as protection from the heat of the sun.

A working horse has no choice, and is resigned to being out in all weathers. A pet dog tends to be fussier, and will probably simply refuse to go out, with or without an umbrella.

Does this mutt look pleased to be out in the wet? And where are his waterproof booties?

153

# PHONE MONOCLE

Unlike many cheap screen magnifiers that easily fall off, fail to stick, or leave dirty residues after you have removed them, the Phone Monocle is marketed as the real deal among a dung heap of no-good pretenders. Made of stretchy rubber, it can be used not just over cellphones, but also home phones and iPods. It acts like a magnifying lens on numbers and messages on the display screen, making them twice their normal size. As if this didn't make it irresistible enough, it also comes in a range of fetching hues, from tropical sea blue, to banana boat yellow. Who could possibly survive without one?

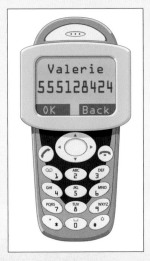

*'Also works wonderfully on insulin pumps.'*

Left: *Phone monocles are advertised as devices that enable you to see who is calling you while you are in your car, without having to take your hands off the wheel and reach for your glasses. But if your close-up eyesight is that bad, should you really be fiddling around with phones while driving?*

The monocle attachment magnifies the screen, but does not extend to the buttons, so don't leave home without those all-important spectacles.

The wrap around design is said to also help stop the 'problem' of the phone sliding around the car.

# PLASTIC HOUSE

Monsanto's House of the Future was touted as a vision of the way we'd all be living by the mid-1980s when it was opened at Disneyland in 1957. As well as the walls and roof being made of plastic, so were sinks, countertops and doors, and even the rugs and chairs. In addition, it was packed with now commonplace 'modern' conveniences, like speakerphones and microwave ovens. The thinking was that plastic lent itself more readily to cheap mass production and efficiency, something every postwar homeowner would be enthusiastic about. Fifty years on, however, people still live in dwellings made of brick and stone, and sit on upholstered chairs of wood and steel because, in the end, a home is not just a matter of technology, but of human expression.

## RATINGS

**STYLE COUNT:** ★★☆☆☆ You probably had to be there to appreciate it

**ORIGINALITY:** ★★★★★

**USEFULNESS:** ★★★☆☆

**NERD APPEAL:** ★★★★☆

**LONGEVITY:** While architects have been adept at incorporating new, lighter materials into building design, the idea of the plastic house slipped quickly off the agenda.

*'Dreams made of plastic.'*

Left: *Over the years there have been several attempts to design a 'house of the future', but they have usually failed because the devotion to technological innovation disregarded human aesthetics.*

The walls of the Monsanto House of the Future were made from bolted together factory-moulded plastic sheets, tested to make sure they would not warp in the heat.

While you might think a plastic house would be flimsy, when they came to demolish the Monsanto house in 1967, a giant wrecking ball used for the job simply bounced off the walls. They ended up having to take it apart bit by bit using hacksaws.

# POGO STICK

There's an unlikely story that Illinois toy designer George Hansburg got his patented idea for a pogo sick from a Burmese peasant farmer on a sales trip abroad. He'd shown him a stick attached to a string that he'd made for his daughter to use to jump through the marshy wastelands that surrounded them, so she could attend church every day. It's a silly story, but then the pogo stick is a silly toy. Hansburg struck gold with it in the 1920s though. For a time, it seemed they were everywhere. People danced on them, and even got married on them. Today you can get ones that enable you to pogo more than 1m (3ft) in the air. But the classic steel pogo sticks fashioned by Hansburg still seem to work the best.

*'The swamp girl's stick that became an international toy.'*

Left: *Pogo sticks were a marvellously green alternative to the gas-guzzling family convertible. Hopefully, it wasn't that far to the local shopping mall.*

The Burmese peasant girl supposedly used her prototype pogo to get to church, but for most folk it was a quick way of getting to the corner store.

Trainee road workers often learnt their skills using a pogo stick.

159

# *PRINTED TOILET PAPER*

$L$ess an invention than an embellishment on an old, trusted and indispensable friend, the printed toilet paper roll is one of those love 'em or hate 'em ideas that really should be thrown straight down the pan. Standardized rolls contain some pretty predictable motifs – dollars or euros, George W. Bush or Osama Bin Laden – none of which are likely to give visitors using your bathroom the impression that you possess a finely tuned sense of the ridiculous. But now customized versions enable you to print anything to order on your toilet paper. You run the risk of revealing a little too much information about yourself, though: images of an ex-girlfriend (you are clearly still bitter); a boss or colleague (conversation had better steer clear of work-related matters). One

Japanese novelist teamed up with a manufacturer to have one of his stories printed on toilet paper. He needn't have bothered. Everybody already knew the plotline stank.

## RATINGS:

STYLE COUNT: ★★★★★

ORIGINALITY: ★★★★★

USEFULNESS: ★★★★★

NERD APPEAL: ★★★★★

LONGEVITY: The sick jokes have already overstayed their welcome.

*'Puts dirty jokes right where they really belong.'*

Left: *It is pointless to get too aerated about the tackiness of printed toilet paper. After all, it's just an attempt to make us laugh.*

Another variation is to print stories on them – a really novel idea.

You can have the paper printed with all sorts of themes, from dollars to politicians, to images of a former lover to the company logo. But the only people who are really having a laugh are the manufacturers.

# REMOTE-CONTROLLED LAWNMOWER

A reliable remote-controlled lawnmower has been the pipe dream of many a lazy suburban lawn jockey over the decades, perspiring on their ugly ride-on machines, resentful at the loss of precious deckchair leisure time. Modern remote-controlled versions claim to be more efficient, both in operation and in fuel consumption, than older models, which were prone to malfunction. But they're still a horrible idea, because they can never be as environmentally friendly as a trusty push mower. Worse, they do nothing to encourage those in the manicured-lawn camp to develop a more symbiotic relationship with nature. If your lawn is so big you can't be bothered to maintain it manually, break it up with beds of shrubs, and plant some trees. That way you can spend more time sunning yourself in your deckchair, and enjoying the hum of the bees and the melody of birdsong while you are at it.

*'Hey... It's mowed my geraniums...'*

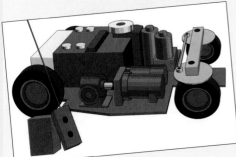

Left: *Early remote-controlled lawnmowers came out in the late 1940s. They essentially looked like ordinary mowers, and the owner operated them by radio control. They could develop minds of their own, however, and run riot across the lawn.*

Modern models are more sophisticated, but they are still nasty, high-level environmental polluters.

Sitting with a remote control frees you from walking up and down the garden with the mower, but not from directing it: more time to sit down, the same amount of time spent mowing the lawn.

163

# ROBOT DOG

On your first encounter, Aibo seemed like the doggy pet from paradise. He responded to his name, obeyed commands to the letter, stood on his hind legs when you patted his head and waved 'hello' with his front paw. When he cocked his leg, he didn't leave a puddle, he wasn't smelly, he didn't shed fur and he incurred no vet bills. Sony, his maker, was certainly convinced that this gleaming metallic mutt was the Next Big Thing.

They'd introduced him in 1999 as the first robot designed for home-

entertainment purposes. Special software enabled Aibo to be raised from puppy stage to adulthood. Sensors and autonomous programs also allowed him to behave like a living creature, and to act on his own judgement. In the end, though, because Aibo was 'only' a robot, owners found he was no substitute for the emotional bond to be achieved with a real animal. In 2006, Sony announced that, sadly, they'd had to put him down.

## RATINGS
STYLE COUNT: ★★★☆☆
ORIGINALITY: ★★★☆☆
USEFULNESS: ☆☆☆☆☆
NERD APPEAL: ★★★★☆
LONGEVITY: Aido reached the grand old age of seven before Sony ruthlessly pulled the plug.

*'He's just eaten my slippers!'*

*Left: Aibo was marketed as the perfect dog: clean, obedient and good with children. But owners soon complained they had difficulties developing an emotional bond with it.*

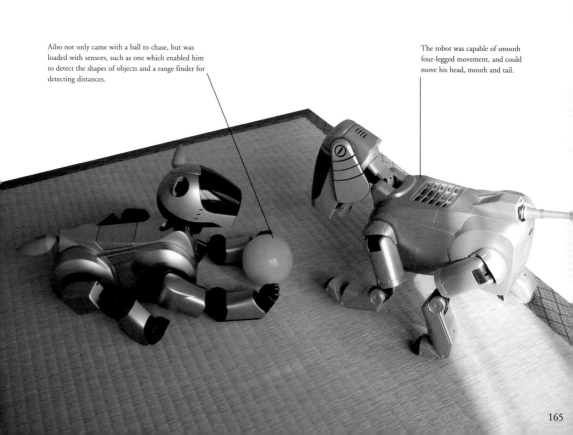

Aibo not only came with a ball to chase, but was loaded with sensors, such as one which enabled him to detect the shapes of objects and a range finder for detecting distances.

The robot was capable of smooth four-legged movement, and could move his head, mouth and tail.

# SEA SHOES

Only Jesus could walk on water, at least until the inventor of sea shoes came along. In reality, the shoes were just a pair of floating skis secured to the user's feet, used in conjunction with water oars with floats attached to the bottom. Movement was possible by sliding the skis forward as you pushed the oars back and forth on the water. They were intended to be a useful means of improving physical fitness, or even engaging in sporting activites. Quite how you could do this while you were never able to do more than gingerly inch yourself forward is difficult to imagine. The sudden twists and turns required to play sport would have been an impossibility.

*'Guaranteed to go down with a splash.'*

Left: *The float on the end of the oar ensured that if you dropped it, it would not sink.*

It's not easy to see why you would want to walk across the water on floating skis, and certainly not wearing a suit.

The bottom surfaces of the skis had rounded ribs to counteract the backward wash of the water, but it's unlikely that forward progress was at all spectacular.

# SELF-MAKING BED

It is no surprise that the inventor of the self-making bed admits to being a lazy male. While the invention would have a particular use for people with difficulties bending their back, it has obvious bachelor pad appeal. The bed linen or duvet cover is spread at the push of a button, and tubes that run along metal rails then smooth it over. Meanwhile straps at the side of the bed will tighten the sheets. Once done, the metal rails are lowered, transforming what could have started out as the untidiest of beds into as neat a finish as your mother ever achieved. As long as it doesn't decide to start making itself while you are still lying in it, here's one 'crazy' invention that really seems to have potential.

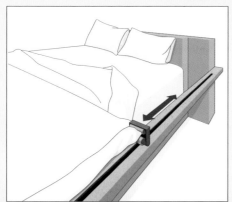

*'Means you can spend an extra five minutes in bed.'*

Left: *There's a lot to be said for a self-making bed, though you have to question whether it will make that weekly changeover of bed linen more of a chore.*

What happens if it decides to start making itself while you are still getting that all-important extra five minutes snooze time?

One drawback is that you'll still have the nuisance of having to move the pillows yourself.

# SILENT ALARM CLOCK

Alarm clocks exert an awful tyranny over domestic living, shrilly rousing us from our blissful slumbers, and disturbing those of our partners too. They are the cause of many an irritable early-morning exchange of opinions. The 'silent' alarm clock dispenses with that annoying racket. Instead, all you get is a gentle vibrating feeling under your ear. The buttons that set the time and the alarm are covered so that they are not accidentally set off when the clock is placed under the pillow. Just in case you miss the vibrations, you can program the clock to send out the conventional alarm sound. There's one

obvious snag here. Few of us sleep in the same position all night. The clock may be strategically placed to vibrate gently under our preferred edge of the pillow when we go to bed, but chances are, by the morning, we will find we have strayed further afield. It's all too likely that most users of these clocks will find themselves having to revert to the wretched noisy alarm method after all.

### 'It will save your relationship!'

Left: *The vibrating silent alarm clock is sold as the alternative to the annoying, relationship-shredding buzz of the conventional clock.*

Just in case you miss the vibrations, there's a back-up audible alarm that can be set, and a built-in light.

It's compact enough to slip comfortably under the pillow. But if you move, you won't feel the buzz.

171

# SQUARE TENNIS BALL

Compared to 'old' sports like golf, soccer, cricket and horse racing, lawn tennis is a relative newcomer to the scene. Its laws, ball, racket and court size specifications only evolved in the late 19th and early 20th centuries. The professional, or 'open,' era didn't get going until the late 60s, and the hard-hitting game we watch today evolved even later than that.

You can have blazing arguments about whether the quality of the tennis has improved. Could it be that the square tennis ball is a sour fan's retort to the baseline bashers, a desperate comment on the stolid footwork of modern players? The ball has all the physical weight and density of a normal round one, fizzes through the air and spins off the racket. It's just that if you don't reach it in mid air, it drops and dies at your feet. But our bitter friend has forgotten that a lightweight form of volleying 'tennis' already exists. It's called badminton, but it's never been a substitute for the real thing.

*'Doubles up as a stress ball for tennis fans.'*

Left: *It's hard to see the square tennis ball as anything other than a joke, but was it meant as an ironic comment on the state of modern professional tennis?*

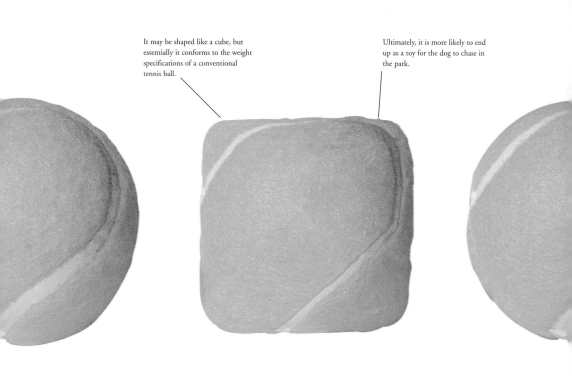

It may be shaped like a cube, but essentially it conforms to the weight specifications of a conventional tennis ball.

Ultimately, it is more likely to end up as a toy for the dog to chase in the park.

# TOILET NIGHT LIGHT

Toilet night lights are small plastic boxes that you attach to the toilet lid. Using an infra-red motion sensor, these clever little things 'see' you approaching in the dark, and bathe the seat in a gentle glowing light. If the seat is down, the light is green, if up, it's red. The issue of whether to leave a seat up or down has been matter of debate between males and females for a long time. The sellers of the night lights will breathlessly tell you that investing in one of these lights will, at a stroke, bring to an end years of domestic squabbling. This may be true, but let's not kid ourselves that it's saved an outbreak of World War III.

*'When you need to glow.'*

Left: *As well as the seat up/seat down conundrum, night lights also overcome the need to switch on noisy bathroom lights in the middle of the night.*

174

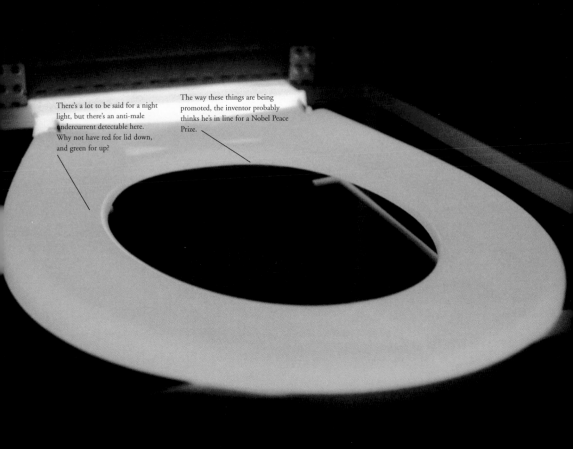

There's a lot to be said for a night light, but there's an anti-male undercurrent detectable here. Why not have red for lid down, and green for up?

The way these things are being promoted, the inventor probably thinks he's in line for a Nobel Peace Prize.

# WEIRD WEAPONS OF WAR

The need to be one step ahead of the enemy has spawned some of the world's most talked about inventions, from bows and arrows to nuclear warheads, but it has also been responsible for some seriously big flops. Admittedly, acoustic listening devices weren't really flops. They were just shown to be obsolete almost immediately by the quickening pace of military technology.

The Sims-Dudley Dynamite Gun certainly was a weapon of terror, often among its own operators as much as the enemy. The Ross rifle was the right weapon in the wrong place, and the Northover Projector was cheap and nasty, and doubtless the makers hoped it would never be required for serious combat. But the sight of the Great Panjandrum firing its payload in all directions on its beach test run, sending official observers scuttling for cover, must rank as one of the most farcical moments of World War II. As for the spectacle of the supposedly supersonic Avrocar spluttering along at below 56km/h (35mph), barely clearing ground level, even its technicians must have laughed (until they remembered the millions of dollars they'd wasted on this exercise in futility).

Left: *The Krummlauf was supposed to enable you to shoot around corners but by the time you'd got it assembled a sniper had probably already blown your head off.*

# ACOUSTIC LISTENING DEVICES

In the 19th century, a ship at sea, lost in the fog and its position unknown, presented a hazard both to itself and other vessels. Early acoustic listening devices, of which the little known topophone of Professor Mayer was one, worked on the idea that a pair of sound-reflecting horns placed either side of the human ear increased the chance of localizing its position. Whether the topophone produced decent results is unrecorded, but the basic idea was good and larger horns in the ensuing decades enabled the tracking of aircraft.

Unfortunately, they still were of limited use in times of war. At the outbreak of World War II in 1939, the so-called 'Listening Ears' might

produce a highly accurate positioning, but fighter planes were also getting faster. By the time you'd charted your reading, the aircraft was already overhead.

## RATINGS

STYLE COUNT: ★★★★★
ORIGINALITY: ★★★★★
USEFULNESS: ★★★☆☆
NERD APPEAL: ★★★★☆
LONGEVITY: Heyday was from around 1914 to the early 1940s, by which time radar technology was replacing it.

*'A foggy reception.'*

Left: *Professor Mayer's topophone, recorded in an engraving of 1880, was one of the earliest audio-location devices. Whether it was much use is unknown, but the basic idea was adapted, and by the 1930s, larger 'listening ears' could detect aircraft in calm conditions about 25km (15 miles) away.*

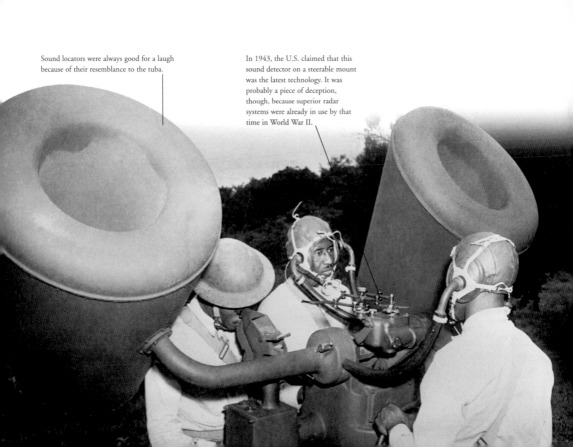

Sound locators were always good for a laugh because of their resemblance to the tuba.

In 1943, the U.S. claimed that this sound detector on a steerable mount was the latest technology. It was probably a piece of deception, though, because superior radar systems were already in use by that time in World War II.

# AVROCAR

The Avrocar was a circular-shaped fighter bomber that looked like a flying saucer, but proved to be about as useful as a frisbee. It was developed by the Canadian company Avro for the U.S. Army and Air Force as a supersonic vertical takeoff aircraft, fired by exhaust from turbojet engines. A turborotor created a cushion of air on which the craft would float at a low altitude, hovering below enemy radar lines, before zooming upward at speeds in excess of 480km/h (300mph). Unfortunately, on its first big run, it only got 1m (3ft) off the ground before it started pitching and rolling uncontrollably. The heat the engine generated was so intense the cockpit controls turned brown, yet the Avrocar couldn't go any faster than 56km/h (35mph). Having already invested millions in the project, the U.S. military cut their losses. However, it wasn't a total failure. The ground cushion principles proved useful in the development of the first hovercraft.

**RATINGS**

STYLE COUNT: ★★☆☆☆ Despite the millions spent, you get the feeling even the Martians might have turned their nose up at this eccentric machine.

ORIGINALITY: ★★★☆☆

USEFULNESS: ★★★☆☆ But not for the purpose for which it was intended.

NERD APPEAL: ★★★★★

LONGEVITY: Almost ten years of painfully expensive experimentation.

*'It looked like a foo fighter, with the emphasis on foo.'*

Left: *In the 1950s, everyone was excited by the idea of supersonic travel. But when the Avrocar trundled along more slowly than a primitive biplane people realized things weren't going to be all that simple.*

You have to hope this pilot wore gloves because the heat the engine generated turned his cockpit instruments brown.

The idea was that the Avrocar could cunningly hover below enemy radar detection levels. But limping along just off the ground wasn't exactly what the designers had in mind.

AV-7055

US AIR FORCE   US ARMY

# CHAUCHAT

Despite being the most widely used machine gun in World War I, the Chauchat was reviled for its haphazard inefficiency. Though its lightness meant that it could be easily slung over a gunner's shoulder and fired from the hip, its defects soon became alarmingly apparent to the French and American soldiers who had to use it. It operated on a long recoil principle, which meant that, after firing, the entire barrel bounced back on

the spring, making accurate aiming a lottery. Continuous firing led to overheating and then the barrel jammed for several minutes until it had cooled off. It was assembled from a variety of reconditioned pieces of metal, so the Chauchat was cheap and easy to produce, but this was scant consolation to a trooper, stuck in the mud of the trenches with a weapon that didn't work.

*'A rapid-fire machine gun that wasn't very rapid.'*

Left: *The Chauchat soon developed a reputation as a useless weapon. Luckily, few reached the front line, and when they did, they were quickly discarded.*

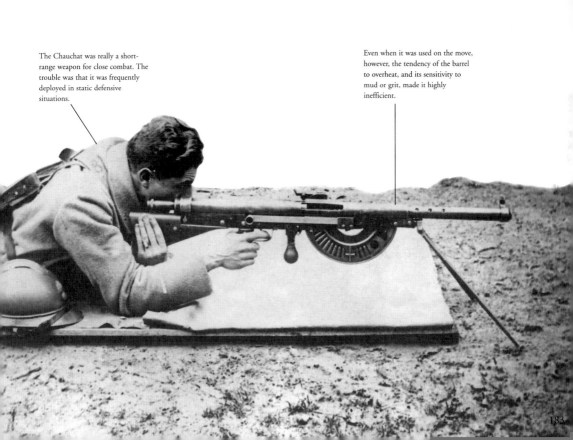

The Chauchat was really a short-range weapon for close combat. The trouble was that it was frequently deployed in static defensive situations.

Even when it was used on the move, however, the tendency of the barrel to overheat, and its sensitivity to mud or grit, made it highly inefficient.

185

# CONVAIR XFY POGO

The Convair XFY Pogo was one of several failed attempts in the early 1950s to devise a combat aircraft that could take off and land vertically, in this case on smaller destroyers or transport ships that could not take conventional aircraft. The Pogo was certainly quick, even at cruising speed knifing along at around 560km/h (350mph). The trouble was that it had no

brakes or spoilers to control airspeed. Landing was also difficult, because it was tricky to judge the rate of descent. When pieces of metal started appearing in the lubricating oil, the Pogo's days were clearly numbered. The last flight of the Pogo took place in 1956, just over two years after it was first conceived.

*'It was supposed to be the last word in sophisticated vertical take-off, but you still needed a rope to climb out.'*

Left: The Pogo actually made more than 280 tethered flights from an airship hangar before its first free flight in 1954. But the development was abandoned once it was recognized that it could never match the performance of contemporary aircraft.

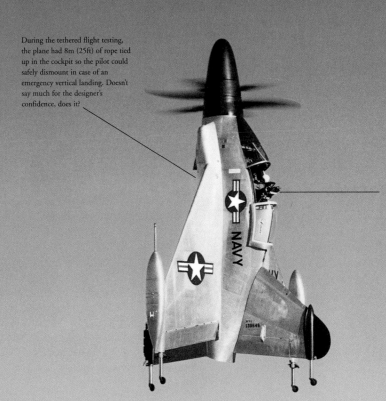

During the tethered flight testing, the plane had 8m (25ft) of rope tied up in the cockpit so the pilot could safely dismount in case of an emergency vertical landing. Doesn't say much for the designer's confidence, does it?

The Pogo was incredibly fast, cruising at around 560km/h (350mph), reaching a top speed of 980km/h (610mph). Strange, then, that nobody thought it was worth providing the pilot with good brakes.

NAVY

138645

# GREAT PANJANDRUM

Of all the outlandish weaponry devised in World War II, the Great Panjandrum must have been among the wildest and weirdest. With D-Day approaching in 1944, the British needed to weaken the German line on the northern coastline of France, and it fell to Lieutenant Nevil Shute Norway (later a successful novelist) to come up with a bright idea.

His Great Panjandrum had two huge wheels, 3m (10ft) in diameter. It was loaded with cordite and, propelled by the force of its rockets at a speed of 96.5km/h (60mph), would roll off a landing craft along the beach. On reaching the defensive wall, it would detonate, leaving a hole large enough to drive tanks though. But on its test run, conducted within full sight of holidaymakers on a British beach, clamps holding the rockets broke and the drum started firing in all directions. After hitting a bump, the Panjandrum, like a blazing Catherine Wheel, changed direction and began heading for a group of official observers and scientists, who fled for their lives. Luckily, the drum then toppled over, and lay smouldering, while it was quietly agreed that, owing to safety concerns, the idea was probably best scrapped.

**RATINGS**
STYLE COUNT: ★★☆☆☆
ORIGINALITY: ★★★★★
USEFULNESS: ☆☆☆☆☆
NERD APPEAL: ★★★★★
LONGEVITY: As an operational weapon it didn't make it past its test run, yet the Panjandrum (minus rockets) was recreated for the 65th anniversary D-Day celebrations of 2009.

*'A real weapon of mass destruction.'*

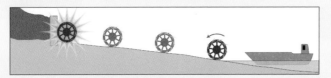

Left: *The Great Panjandrum was a massive, rocket-powered wheel designed to blast a hole in the German seafront line of northern France. But it never made it beyond its final testing, when it ran amok on a beach and sent assembled observers diving for cover.*

The Great Panjandrum comprised two huge wooden wheels, joined by a central drum carrying the explosives. Sets of cordite rockets attached to each wheel created the propulsion.

The idea was that when loaded with 1815kg (4000lb) of explosives, the Panjandrum, upon reaching a speed of 96.5km/h (60mph), could simply crash through anything in its path.

# KRUMMLAUF

It may have looked ridiculous, but the Krummlauf was the result of a spirited effort by Germany toward the end of World War II to give its beleaguered troops the capacity to shoot around corners. Varying models came with a barrel attachment curved at angles of anything from between 90–30 degrees, with a periscope on the end to enable the user to take aim. Predictably, however, the Krummlauf took forever to assemble. Even worse, there were holes in the side of the barrel that were meant to allow cooling gases to flow out. When they became blocked, the entire gun exploded.

**RATINGS:**
STYLE COUNT: ★★★★★
ORIGINALITY: ★★★★★
USEFULNESS: ★★★★★
NERD APPEAL: ★★★★★
LONGEVITY: A short-lived experiment

*'What happened when it was a left-handed corner?'*

**Left:** *The Krummlauf's most extreme version had a barrel curved at 90 degrees. It looked more like something you cut hay with.*

By the end of World War II, Germany needed some inspired thinking, but the Krummlauf wasn't it.

A periscope helped you take aim, but it took an age to put the Krummlauf together, and the weight must have made aiming awkward.

# NORTHOVER PROJECTOR

Facing the threat of German invasion early in World War II, the British Army set up a local volunteer force known as the Home Guard to defend the country. The Northover Projector was to be a cheap and easy way of arming them with anti-tank weaponry. It certainly looked elementary – the Northover was essentially a steel pipe with a basic breech mechanism on the end, standing on a tripod. The idea was that it would propel bottle mortars filled with phosphorus. However, although the sighting range was adequate up to 100m (330ft), it was hopeless beyond it. In any case, the so-called self-igniting grenades proved useless, the glass bottles often breaking up on firing, with nasty consequences for the gun crew.

*'It looked like a toy cannon, and was just about as effective.'*

Left: *The self-igniting phosphorus bottle 'grenades' were loathed by the soldiers because they often broke up on firing.*

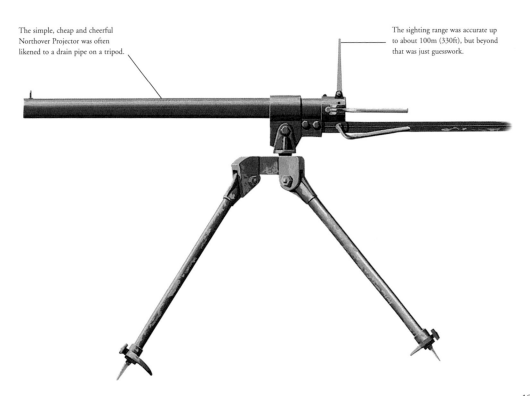

The simple, cheap and cheerful Northover Projector was often likened to a drain pipe on a tripod.

The sighting range was accurate up to about 100m (330ft), but beyond that was just guesswork.

# PERCIVAL P.74

The Percival P.74 helicopter's big advantage was to be that it had tip-jet powered rotors, thus overcoming the air drag, or torque, associated with other helicopters. It never got off the ground. The rotors were powered by gas generators, which sent compressed air through hot, noisy pipes that ran up the cabin walls to the blades. The test pilot, noting that there was no entrance door or escape hatch near the cockpit, and that the only exit was to the rear of the fuselage, said the whole affair had been designed without any input from a pilot. A consultant designer was said to have used the wrong formula for calculating lift. Come test day in 1956, the P.74 clearly wasn't going anywhere. The compressors simply didn't generate sufficient energy to get it off the ground, and it was last seen being humiliatingly towed off the airfield.

**RATINGS**

STYLE COUNT: ★★★★★
ORIGINALITY: ★★★☆☆
USEFULNESS: ☆☆☆☆☆
NERD APPEAL: ★★★★★
LONGEVITY: It simply never took off.

*'Even by helicopter standards, it was ugly.'*

Left: *The P.74 was supposed to usher in a new era of smooth helicopter flight, but its first trip to the airfield proved to be a short detour on the way to the scrapheap.*

With a fuselage shaped like a teardrop, and a cabin running its full length, it wasn't a pretty sight. That might not have mattered so much if had it at least managed to become airborne.

While the tip-jet rotor system was supposed to overcome torque, the gas generators that powered it made for a noisy, sweltering cabin.

XK889

# ROSS RIFLE

The Ross rifle, named after its developer, the Canadian aristocrat Sir Charles Ross, had acquired a fine reputation on the target range in the years leading up to World War I. Once handed to Canadian troops in the trenches, however, its shortcomings were painfully apparent. Faced with

rapid-fire conflict, not only did its sensitivity to dirt and mucky ammunition cause it to jam, it had a complicated bolt system which, if improperly reassembled after cleaning, could result in the gun backfiring. When the rifle was fired, the bayonet attachment often fell off. Stubbornly, the Canadian high command stuck by the gun until 1916 before ordering the deployment of the more reliable Lee-Enfield, by which time Ross had pocketed a fortune.

## RATINGS

STYLE COUNT: ★★★☆☆

ORIGINALITY: ★★★☆☆

USEFULNESS: ★★★★☆ A perfectly good gun when used for recreational purposes, it was just the wrong weapon for the battlefield.

NERD APPEAL: ★★★★☆

LONGEVITY: Overcame its World War I flop to become a trusted rifle of choice among deer-stalking types.

*'The gun that got stuck in the mud.'*

Left: The Ross rifle was a lot of fun when used for hunting deer, but soldiers weren't amused when they had to rely on them in the trenches.

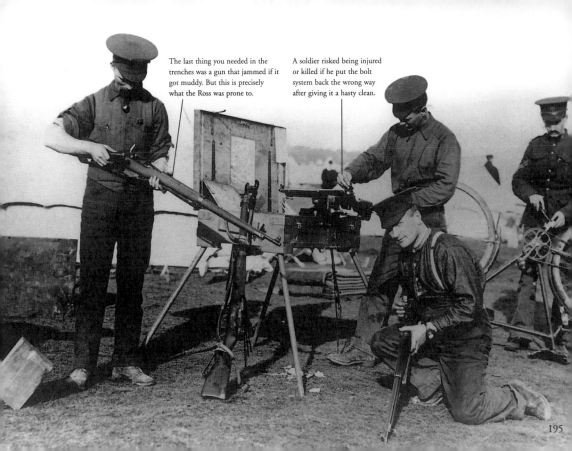

The last thing you needed in the trenches was a gun that jammed if it got muddy. But this is precisely what the Ross was prone to.

A soldier risked being injured or killed if he put the bolt system back the wrong way after giving it a hasty clean.

# SHORT SEAMEW

The buildup of Soviet naval power in the 1950s had NATO military planners anxiously seeking new defensive weaponry of their own. The Short Seamew was not one of the smartest answers. It was a light anti-submarine reconnaissance aircraft designed to operate with escort carriers of NATO forces. Its main assets were to be simplicity of design and cheapness of construction. But bomb and depth charges crammed in under the fuselage and further armaments attached beneath the wings made it a nightmare to handle. This was a serious failing, given it was expected to patrol flights of long duration. By the time a further prototype had crashed at an air display in 1956, killing its pilot, the British Royal Air Force had long since called off the project.

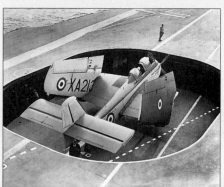

*'A camel amongst racehorses.'*

Left: *The Soviets were pouring vast sums into military spending in the 1950s. Cashed-strapped western governments responded in kind, but spent as little money as possible. The cheap but inefficient Short Seamew was taking budget slashing a step too far.*

The Seamew was loaded with weaponry that not only gave the fuselage its deep, rather odd appearance, but put the pilot and observer uncomfortably high in the cockpit. It didn't make for easy handling.

They were designed to be transported on NATO ships. But while the Seamew was easily stored on the boat, it was less than agile in the air.

Short Seamew

197

# SIMS-DUDLEY DYNAMITE GUN

Using compressed air to fire its projectiles, the Sims-Dudley was deployed by Theodore Roosevelt and his Rough Riders during the Spanish-American War in 1897. It was useful as a shock weapon, able to blow up targets more than 460m (500 yards) away, but its defects outweighed its advantages. Because the gun cylinder had to be aimed high, projectiles were liable to blown off target in high winds, limiting accuracy. The projectiles were sensitive to freezing, and often simply failed to detonate. After every few shots, the gun mechanism would jam, forcing soldiers to waste a couple of hours repairing it. Further developments in high explosives and inherent drawbacks with all dynamite guns led the U.S. government to deem them 'not acceptable' by 1900.

**RATINGS**

| | |
|---|---|
| STYLE COUNT: | ★★☆☆☆ |
| ORIGINALITY: | ★★☆☆☆ |
| USEFULNESS: | ★★☆☆☆ |
| NERD APPEAL: | ★★★★☆ |
| LONGEVITY: | Fired in combat for the first time in 1897, but had been wheeled off into retirement before the century was out. |

*'The dynamite gun that sounded like a powder puff.'*

Left: *The name 'dynamite gun' was actually a misnomer for the Sims-Dudley, because the mortars it fired were actually comprised of a nitro-based gelatin.*

When the Sims-Dudley was first fired on the battlefield, no one wanted to stand next to it. But because it used compressed air, the noise just sounded like a loud cough.

It could be devastating, over a short range, that is, and if you managed to get it on target.

# SNECMA COLEOPTERE

The SNECMA Coleoptere looked like something out of a classic 1950s sci-fi movie, but was actually supposed to be a high-speed airfighter. It was designed by French technicians to take off vertically, which meant that it could use any piece of tarmac or field to launch from or to land in. Its air speed (up to 800km/h/500mph) meant it would be quickly deployed, a key asset during the Cold War when Western European countries could expect little warning of an attack from the East. Once in the air it was to switch to

the horizontal position of a conventional aircraft. But on its maiden flight in 1959, its instability was dangerously apparent. As it screamed to a height of 152m (500ft), the pilot lost control as he tried to set the craft to horizontal. He was seriously injured in ejecting from the plane, which itself was destroyed as it crashed to the ground in a blaze of fire. Further development was abandoned.

*'It went up fast, but came down even faster.'*

Left: *During the Cold War, military technicians needed to develop weaponry able to respond to the danger of rapid air or nuclear attack without much warning. But the only people who were endangered by the Coleoptere were their own personnel.*

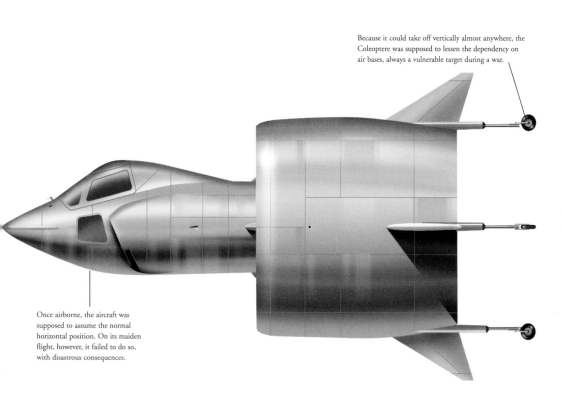

Because it could take off vertically almost anywhere, the Coleoptere was supposed to lessen the dependency on air bases, always a vulnerable target during a war.

Once airborne, the aircraft was supposed to assume the normal horizontal position. On its maiden flight, however, it failed to do so, with disastrous consequences.

# STICKY BOMBS

The sticky bomb, or Grenade, Anti-Tank no. 74, was a makeshift grenade hurriedly devised to make good a shortfall in British anti-tank weaponry in World War II. Field guns were ineffective against tanks, and conventional hand grenades simply bounced off of them and exploded in the air. The sticky bomb was an explosive contained within a glass sphere which was held in a woollen stocking coated in extremely adhesive birdlime. The soldier would either hurl it at a tank, or, if they could get close enough, attach it to the body of the vehicle, releasing the explosive lever at the same time. Unfortunately, the resin was just as likely to stick to the user's clothing, and several soldiers lost their lives as the grenades detonated. Ultimately they were designated as a weapon of last resort for the Home Guard, though sticky bombs were used by British and Commonwealth fighters in the campaign for North Africa.

*'It took "sticking to the task" to new levels.'*

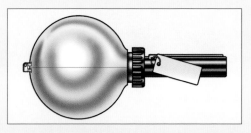

Left: *Sticky bombs seemed like a cheap answer to the threat of a German invasion early in World War II. But they were so sticky they were more likely to blow up their users than the enemy.*

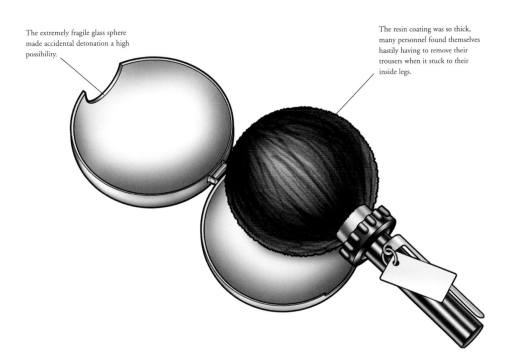

The extremely fragile glass sphere made accidental detonation a high possibility.

The resin coating was so thick, many personnel found themselves hastily having to remove their trousers when it stuck to their inside legs.

# V3 LONDON GUN

The most impressive thing you could say about the V3 was it was the largest gun ever built. The V3, or London Gun, was built with the intention of pounding the capital city with a continuous volley of shells. It was to be an emplacement of around 50 guns, supplied by a sunken barrel 140m (460ft) long, situated across the English Channel on the cliffs above Calais. But this would-be weapon of mass destruction was never put into full action because its concrete bunkers were destroyed during the 617 Squadron 'Dam Buster' raids of 1943. Even if the bombers had not been successful, a few test shots had already suggested that the V3 shells were too feeble to have ever reached London.

*'A big gun, with a puny firepower.'*

Left: *Though the concrete bunker had a 45.5m (15ft) thick protective dome, it crumbled under the massive bombs dropped by the 617 Squadron.*

The handling of the V3 was an example of Hitler's hubris. He ignored his military advisors, who argued that the V3's high coastal position would leave it vulnerable to enemy bombers. The Allies soon destroyed the emplacement.

The V3 guns were supposed to be capable of firing 140kg (308lb) shells over a 165-km (102-mile) range, to strike London. It was another wildly improbably Nazi plan.

# TRANSPORTATION INNOVATION

It's only fair that Transportation Innovation, like every other section in the book, contains its share of tongue-in-cheek gimmicks. For example, Carl Fisher's balloon car (though people were actually taken in by it on the day it 'flew') was just a stunt to promote a new car showroom. And the conference bike was always destined to appeal only to rather sad attention-seekers.

Inventions are at their most fascinating when they earnestly aim high but end up as resounding technological failures. Can you imagine being there to see Sir Henry Bessemer's luxurious anti-sea sickness boat the day it crashed into the pier at Calais? Or how about the unnerving experience of being a passenger in the carriage of the curious Daddy Long Legs tottering high above the crashing waves the day it stalled on a stormy day on Brighton seafront? It's hard to believe now, but the Otto Cycle actually boasted a legion of precariously seated advocates in its heyday. Could you make the same claims for the Unicycle today? Here, too, you can read about the ill-fated Sinclair C5, a widely trumpeted (and crushingly disastrous) experiment in transport innovation.

Left: *The anti-sea sickness boat offered many creature comforts to passengers, but safe steering wasn't one of them.*

# AERO BICYCLE

Everyone knows the story of the Wright brothers and the birth of aviation, but at the same time many diehards were still wedded to the idea of human-powered, as opposed to engine-powered, flight. Aviettes looked like biplanes, but had bicycles attached. Pedal power was the substitute for the propeller, and there were no other aerodynamic forces to call upon apart from the extremely wide wings. Naturally enough, nobody

had much luck flying the aviettes, though one man did manage to get the thing airborne, even if for no more than the distance a long distance jumper could leap.

*'Backflips in the sky.'*

Left: *The possibilities of human-powered flight remained so fascinating, even sharp-suited bank workers would try it out during lunch breaks. Though they had to be careful their ties didn't get caught in the propeller.*

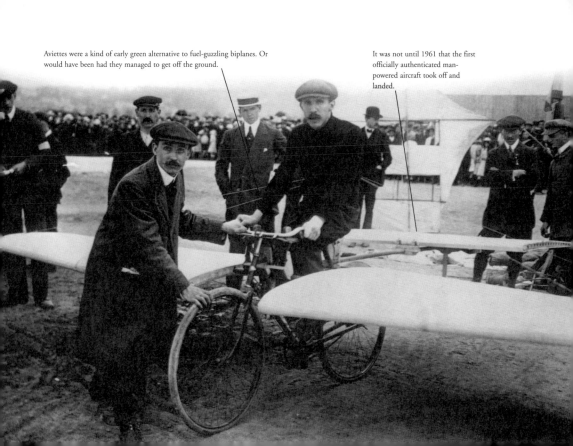

Aviettes were a kind of early green alternative to fuel-guzzling biplanes. Or would have been had they managed to get off the ground.

It was not until 1961 that the first officially authenticated man-powered aircraft took off and landed.

# AEROCAR

On the face of it, the Aerocar still seems like a brilliant idea. As roads get more congested by the year, and air travel more of an ordeal, what could be better than a car that turns into a private airplane at the press of a few buttons? But the Aerocar, designed and built by Moulton Taylor in Washington in 1949, never reached production stage. It wasn't that it was difficult to set up. Taylor once drove his prototype into a television studio and transformed it from road vehicle to plane in just over three minutes, thanks to the wings, tail and rear-mounted propeller being stored in a trailer towed behind the car. It was a nifty little mover, and Taylor proudly

recounted how he'd received a speeding ticket while driving it to a car show. But while the panels were made of fiberglass to reduce weight, it failed to find a manufacturer for the same reason all subsequent attempts to create a flying car failed to reach production. The technology involved in creating a dual-purpose vehicle simply made it too heavy to get airborne.

*'If James Bond movies had low budgets, they'd probably have come up with something like this.'*

Left: *With more backing from a major car manufacturer like Ford, Moulton Taylor might have had more chance to develop his flying car prototypes further. But company executives were loathe to risk reputations on what might seem a hare-brained idea.*

Unfortunately, once up in the air, the road vehicle elements created too much dead weight for the car to stay aloft for long.

The fact that the Aerocar's flight components were towed in a trailer didn't compromise its roadworthiness.

# AUSTIN MAXI

Alec Issigonis had a god-like reputation in the European motor industry as the brain behind iconic cars like the Morris Minor and the Mini, but he hit a wall with the Austin Maxi. Despite being launched to a fanfare of publicity in 1969, the car was soon being attacked for its lack of engine power, terrible gear changes and tendency to break down. Its rather staid appearance lacked sex appeal and, over time, it proved to be something of a rust bucket. For all that, it had a lot going for it. It was a roomy and comfortable five-door hatchback. For a family car it was competitvely priced and not overly expensive to run. Production sputtered on through the 1970s, undergoing various cosmetic adaptations, but the lowly reputation it had acquired in those early years proved impossible to live down.

## RATINGS

| | |
|---|---|
| STYLE COUNT: | ★★★☆☆ |
| ORIGINALITY: | ★★★★☆ |
| USEFULNESS: | ★★★☆☆ |
| NERD APPEAL: | ★★★★☆ |
| LONGEVITY: | A dozen years – not bad going for the financially stretched, strike-ridden British car industry of the 1970s. |

*'Room enough for the tallest Guardsman.'*

*Left: The Maxi had a solid, functional look but buyers of family cars by the turn of the 1970s were after something a little more racy.*

The Maxi had five doors and fold-down seats, which put it miles ahead of its rivals.

While much was made of the car's roominess and adaptability, more attention to the gear mechanics and premature rusting might have helped restore its fortunes.

# BALLOON CAR

It was one thing to dream of flying an airplane around 1900, but what about flying a car? Some years earlier, one intrepid soul had actually attempted to fly a gliding horse cart which, needless to say, never got off the ground. Other efforts followed, but it's fair to say most were no more than

stunts, such as that pulled off by Carl Fisher, entrepreneur and salesman, in the first decade of that brave new century. In an effort to promote his new car showroom, Fisher attached a hot air balloon to a white Stoddard-Dayton automobile, and flew over downtown Indianapolis into the surrounding countryside. He'd actually removed the engine to lighten the load, but the crowd of thousands who gathered to watch didn't know that. When he roared back into town in an identical model he'd strategically hidden, he received a hero's welcome.

*'The first car to fly over Indianapolis – make the Stoddard Dayton your first car too!'*

Left: *It looked like Carl Fisher had designed a flying car, but he had removed the engine.*

Inevitably, Fisher's stunt was copied by others, such as this gentleman, who must have been in the flour business.

VICTOR EDISON'S
SELF RAISING
AIR CAR

Flying a car wasn't Fisher's only stunt. He once pushed a car over the edge of a building and then drove off in it to demonstrate its durability.

215

# BAZIN'S ROLLER BOAT

When the blueprints of Ernest Bavin's roller boat were unveiled in 1895, observers nodded sagely and said the boat would probably be capable of reaching thirty-two knots, faster than any liner of the period. Men had been trying to make a good roller ship (a boat propelled by wheels) for years without success. Bazin calculated that, by lifting the hull clear above the water, the giant wheels (each one of the three pairs was 30m (89ft) in diameter and 3m (10ft) thick) would be left free to obtain maximum propulsion.

In the end the 300-ton beast turned out to be a big white elephant. Launched on the English Channel in 1896, the wheels churned up so much water they caused a braking effect, and the vessel stalled. Bazin, ever the optimist, would go to his grave claiming a solution was just around the corner.

*'A ship that created big waves but went nowhere.'*

Left: *Lifting the hull clean above the water was Ernest Bazin's solution to getting a wheel-powered boat to work. The deck certainly lacked nothing in terms of an ornate wrought iron finish.*

216

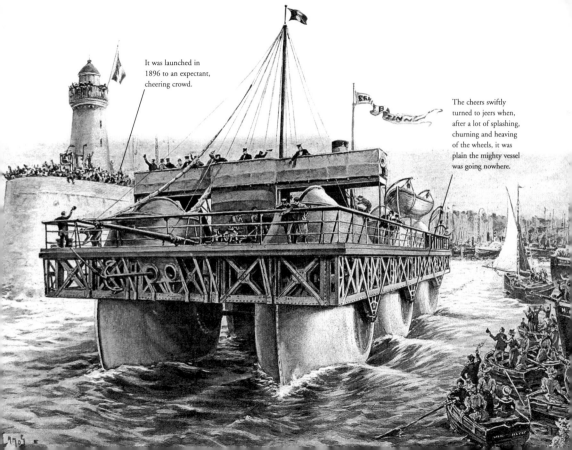

It was launched in 1896 to an expectant, cheering crowd.

The cheers swiftly turned to jeers when, after a lot of splashing, churning and heaving of the wheels, it was plain the mighty vessel was going nowhere.

# BENDY BUS

T wice the length of a regular public transport vehicle, bendy buses were designed as a means of alleviating the increasingly congested roads of towns and cities. With at least three sets of doors, they permit quicker passenger access, and they have a larger carrier capacity than even double-decker buses. However, to facilitate their greater length, they have an extra set of wheels and a joint (the 'bend') about halfway along, which makes them

incredibly cumbersome and slow to steer. It seems they actually cause more traffic jams and accidents, partly because drivers of other vehicles are desperate not to get stuck behind them, and cyclists and pedestrians have difficulty seeing around them to gain a view of incoming traffic. There have also been several incidents of the buses overheating and catching fire. There has probably has never been a form of public transport as unloved as the wretched bendy bus.

RATINGS

| | |
|---|---|
| **STYLE COUNT:** | ★★★★★ |
| **ORIGINALITY:** | ★★★☆☆ |
| **USEFULNESS:** | ★★★☆☆ |
| **NERD APPEAL:** | ★★★★★ (Just wait until they are finally taken out of service. The defenders will be out in force.) |
| **LONGEVITY:** | Way too long. |

*'The only prize it ever won was Worst Design Ever.'*

*Left: They seemed like a way to quicken up the flow of traffic on streets, but because they are so cumbersome, bendy buses only exacerbate the problem.*

218

The great length of around 18m (60ft) compared to 10–12m (35–40ft) for other types of buses may mean you can get more people on board, but also makes them a nightmare for other vehicles, cyclists and pedestrians.

The buses allow quicker and higher levels of passenger access, and are supposed to be more convenient for the disabled or elderly. But most people say the seats are uncomfortable, that is, if you are even able to get one.

# SS BESSEMER ANTI-SEA SICKNESS BOAT

Sir Henry Bessemer amassed a personal fortune in the 19th century through his many inventions, but his anti-sea sickness boat was not a notable contributor. Having suffered a severe bout of sea-sickness on a voyage across the English Channel in 1868, Bessemer turned his restless mind to devising a remedy. He designed a steamship so long that it would ride four or five waves without pitching fore and aft, while the passenger cabin was mounted on axles with a balancing weight underneath, so that it would remain horizontal, however much the vessel rolled on the ocean. The

interior of the saloon was beautifully fitted with morocco-covered seats, carved oak divisions, spiral columns and gilded panels.

However, the immense central weight made the vessel difficult to steer. In May 1875, the SS *Bessemer* sailed out of Dover port on her maiden voyage but, attempting to enter the docks at Calais, demolished the pier. Bessemer accepted failure, costly as it was, and the ship never put to sea again.

*'The anti-sea sickness boat that was an equally sickening flop.'*

Left: *Sir Henry Bessemer himself later described the unhappy culmination of SS Bessemer's maiden voyage in ebullient style: 'Crash she went along the pier side, knocking down the huge timbers like so many ninepins!'*

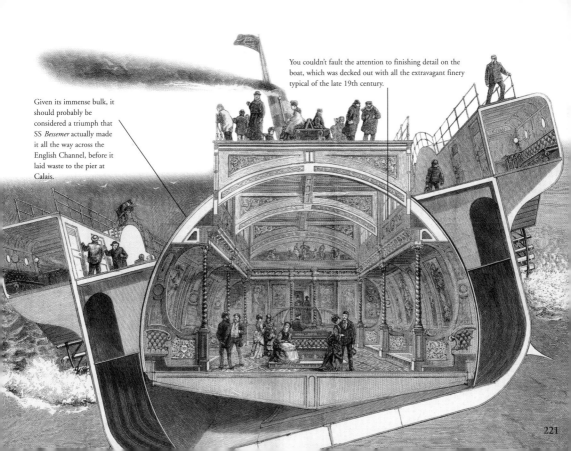

Given its immense bulk, it should probably be considered a triumph that SS *Bessemer* actually made it all the way across the English Channel, before it laid waste to the pier at Calais.

You couldn't fault the attention to finishing detail on the boat, which was decked out with all the extravagant finery typical of the late 19th century.

221

# BLACKBURN SCOUT

I f the British Admiralty thought they needed to repel the threat of the strange-looking Zeppelin bombers with an equally strange-looking fighter aircraft, they certainly would have had it if they'd ever gone ahead with the Blackburn, or AD, Scout. In fact, only four prototypes, two by the Blackburn aircraft company, were ever made, but it must have had official brows furrowing the moment they saw it.

The Scout, or Sparrow, as it was commonly known, was a biplane which, most unusually, had the fuselage attached to the upper wing. While this high position gave the pilot an excellent field of vision for spotting low-flying airships, the slender undercarriage did nothing for stability and the plane had an unnerving tendency to tip forward. It was also quickly apparent that it was too fragile to lift the weight of the required weaponry. After test pilots gave it a unanimous thumbs down all four prototypes were scrapped.

*'I say, chaps, it's desperately lonely up here.'*

Left: *The unprecedented threat posed by German Zeppelin airships may have required an imaginative response by British military eggheads, but breaking with convention and mounting the fuselage on the top wing of a biplane, as the designers of the Scout did, was not the answer.*

The plane was supposed to be loaded with a two-pounder recoilless Davis Gun, but this frail structure was never going to be capable of carrying such weighty weaponry.

A tailplane was attached by four slender booms, whose skids were supposed to stop overbalancing. But the tailplane itself was as large as the upper wing.

223

# BOND BUG

The Bond Bug was the result of a merger in 1970 between two British small car manufacturers, Reliant and the struggling Bond. The Bug, a futuristically wedge-shaped three wheeler, looked like something Luke Skywalker might have driven. It was only available in lurid orange, which supposedly stayed bright even when the car was getting dusty. Whereas the average age of the Reliant Robin driver was supposed to be over sixty, the Bug was aimed at the late-teenage to early twenties age bracket. In terms of speed it could hold its own very well against small

four-wheeled cars of the period, but it was always too much of an oddity to ever catch on, especially with younger drivers.

*'Would have looked better if they'd been pink.'*

Left: *The Bond Bug enjoyed only a modest production life, but its appearance was somewhat ahead of its time. Perhaps this explains why many are still to be seen on the road today.*

The idea behind the Bug was that it would be a sporty three-wheeler, appealing to a younger generation, but a top speed of 125km/h (78mph) wasn't going to get many people too excited.

Getting into the car was reminiscent of getting into the cockpit of a small plane. You pulled the lift-up front canopy and climbed in.

DUV 91J

# *CAPRONI CA.6*

The Italian count Gianni Caproni had an estimable reputation as a builder of fighter bombers in World War I, but he risked tossing it to the four winds in 1921 with his grandiose Ca.60 flying boat, designed to carry over one hundred people across the Atlantic from Italy to the U.S.

The Ca.60 had three sets of giant triple wings amounting to twice the wing area of a B-52 bomber, and had eight 298kW (400hp) engines. The pilot sat in an open cockpit, but the passenger cabin had more window glazing than an airliner. However, upon making a short flight over the Italian Lake Maggiore it rose about 18m (60ft) in the air before abruptly nosediving into the water. Fortunately, no one was killed, but a mysterious fire that

broke out while the crumpled structure was undergoing repairs destroyed Caproni's flying-boat dream forever.

*'A houseboat on wings.'*

Left: *The Ca.60 was a spectacular affair, but was actually merely a prototype for an even larger, more ambitious plane that would have had capacity for over 150 passengers.*

226

For all the grandeur of the concept, Caproni had merely reused wings from some old wartime bombers, which he bolted onto the flying boat structure. Lack of tail surfaces made the craft inherently unbalanced.

The eight 298kW (400hp) engines gave it ten times the amount of power of an average passenger aircraft of the time.

# CAR ALARM SYSTEM

It's unlikely that there has ever been an invention that has tempted more otherwise, mild, sane, law-abiding people into launching a malevolent assault on another person's property than a car alarm system in full cry. What is more guaranteed to raise the heart rate to dangerous levels than its insistent scream on a steaming hot day in the city, or to reduce a night's slumbers into a restless bout of sleepless tossing and turning?

Car alarms were designed to warn off potential thieves thanks to the installation of an electronic device that sets off its sound with accompanying flashing lights. But because the sirens are so easily triggered, no one pays any attention to them, and criminologists have confirmed what everyone else already knows – that car alarms are completely useless as a deterrent.

*'91% of New Yorkers say car alarms diminish their quality of life. And only 5% say that the sound of one would cause them to take action against possible theft.'*

Left: *Car alarm systems were supposed to deter criminals, but because they are so easily set off by the vibration of passing trucks, or even the bass level of a sound system, they are ineffective.*

More car manufacturers are fitting effective 'silent' alarm systems, which haven't arrived a moment too soon.

The systems can be controlled by the owners using a remote key fob. But if they unaware of the alert, it's hell for everyone else.

# CAR ROOF TENT

As the world's economy slowed to a crawl toward the end of the first decade of the new century, if there was one group of people rubbing their hands in glee, it was those in the car roof tent business. A lot of folks, so their reasoning went, can't even afford a motorhome or trailer, let alone a holiday abroad, so they assumed their product would be in big demand.

It doesn't matter how small your car is, as long as you are prepared to spend a night sleeping on its roof, and the car roof tent guys will certainly be happy to help you. Most will spin a line about how roof tent technology has 'greatly improved' since it was first devised by the Italians in the late 1950s. So? Maybe it has. But if the winds get up, and the rain starts lashing down,

it's a hardy vacationer who's prepared to tough it out up there, whether under a basic canvas sheet, or something you might take to the Antarctic.

*'Cheap – and nasty.'*

Left: *In the swinging 60s, it was considered quite cool to pitch up for the night under a car roof tent. However, not every supplier was considerate enough to provide a ladder.*

Preferably, it was just you going on vacation, because there wasn't going to be much room for the family.

All you had to do was fold it up again at the end of the night and away you went.

# CONFERENCE BIKE

The conference bike is one of those contraptions that was clearly designed by someone who thought he was being funny, when all they'd really achieved was creating something immensely stupid and irritating. In essence, the bike is a circular, multi-rider cycle that is supposed to make passersby as well as riders, feel 'cheerful'. Only one of these presumably go-getting riders steers, while the others merely pedal.

No word yet that wellness coordinators have adopted the thing as a tool for improving social interaction in the office, but it's probably too early to be uncrossing those fingers.

*'Fun is the one thing you won't have.'*

Left: *People have had some ridiculous ideas for bicycles over the decades, but this one stands out as a real stinker.*

232

The cycle is supposed to combine the social aspect of sitting in the company of a group of friends with the pleasures of being outdoors. Never heard of sitting in the park?

Riders of conference bikes seem to think that when pedestrians gape at them in disbelief, it's because they're amused. It's more likely they are just feeling sorry for them.

# DOUBLE HANDLEBARRED BICYCLE

The double handle-barred bike, or social bicycle, has been around for almost as long as the tandem. In the early days you could see the appeal. It gave young courting couples a chance to escape the disapproving glare of the parental setting and enjoy a bit of knee-to-knee contact, while ostensibly on an 'innocent' ride in the country. Who, though, would need such a thing today? The pairings on these bikes look like the very picture of

discomfort. Not the least of the problems you could anticipate are potential weight disparities that cause lopsided balance issues, and poor pedal-power coordination. Honestly, how close to your 'buddy' do you really need to be?

**RATINGS**

STYLE COUNT: ★★★★★

ORIGINALITY: ★★★★★

USEFULNESS: ★★★★★

NERD APPEAL: ★★★★★

LONGEVITY: A paltry level of curiosity value has seen them last for nigh on a century.

*'Keep your friends close… but this close?'*

Left: *The double handle-barred bike just might have had some appeal a century ago, but few furtive courting couples would resort to it today. Meanwhile, short of sharing top-secret information, it's hard to see the point for a platonic duo.*

Steering is only done by one of the riders, with the second pair of handlebars being non-rotatable.

The bike operates by a pair of pedals on either side of the crankshaft, meaning they are both connected to the rear wheel.

235

# EXCALIBUR SSK

It looked like a gaudy vintage roadster, but was powered by a blistering Chevrolet Corvette engine. Enthusiasts deplored the fakery of the concept, but connoisseurs of the unusual lapped it up, and it spawned a new line in 'replica' cars. The Excalibur SSK certainly divided opinion. It was launched in 1964, modelled on a sporty 1930s classic, the Mercedes SSK. A two seater, it had plush leather upholstery, freestanding headlights and decorative side-exit exhausts. It was more expensive than a Cadillac, but a bad power-to-weight ratio made it tough to drive, and confirmed its status in the eyes of its detractors as a mere neoclassical fraud.

**RATINGS**
STYLE COUNT: ★★★★☆
ORIGINALITY: ★☆☆☆☆
USEFULNESS: ★★☆☆☆
NERD APPEAL: ★★★★☆
LONGEVITY: 15 expensive years.

*'And you won't even burn yourself on the (fake) exhaust pipes.'*

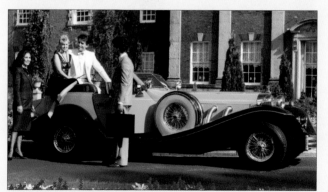

Left: *If the Excalibur SSK was expensive, a deluxe version complete with swooping front fenders and running boards retailed at a (for the time) eye-watering $8000. No wonder they were deemed to be nothing more than rich men's playthings.*

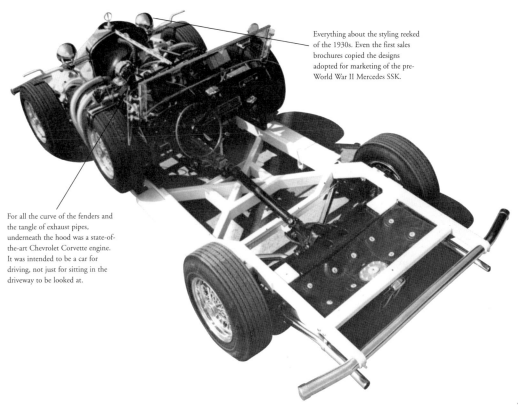

Everything about the styling reeked of the 1930s. Even the first sales brochures copied the designs adopted for marketing of the pre-World War II Mercedes SSK.

For all the curve of the fenders and the tangle of exhaust pipes, underneath the hood was a state-of-the-art Chevrolet Corvette engine. It was intended to be a car for driving, not just for sitting in the driveway to be looked at.

# FORD MUSTANG MK II

If the original Ford Mustang had an air of danger and menace, the MK II, introduced in 1973, was disappointingly demure. It wasn't just that, in an era of fuel rationing and increasingly stringent restrictions on emissions, its floor size and engine power were much reduced. It was simply that its design was so clunkily staid in comparison to the tarmac-shredding Mustang of popular imagination. To be fair, it actually sold very well, and the president of the Ford Motor Company affectionately dubbed it his 'Little Jewel'. But hoots of derision from style connoisseurs ensured that the cult status of its swinging 60s ancestor passed it by completely.

*'If you got 145km/h (90mph) out of it, you threw a party.'*

Left: *The MK II traded on a sharp brand reputation as a sporty classic. But its prim design made it look more like the car you took granny shopping in.*

238

Improved suspension did make for a smoother ride and better handling, but the interior was mocked for its 'wood grain appointments' and 'chintzy décor'.

It was fitted up with a catalytic converter. Good for the environment, perhaps, but a serious knock to its reputation as a speedster.

# HORSE-DRAWN CAR

Most inventors are driven by the idea of progress, but with the horse-drawn car it's possible to detect a more wistful, backward-looking intent that borders on nostalgia. Outside of Amish communities, horse power hasn't been much used as a serious means of getting around in the West for decades, but the idea of it still packs a considerable emotional clout. The musical clip-clop of hooves on the road and the whiff of manure call us back to our rural roots. Having the car merely attached to the back like some rusty old plow puts the modern giant of the highways that usurped them right back in its place. Totally ludicrous, of course, but good to see them around, just the same.

**RATINGS**

| | |
|---|---|
| STYLE COUNT: | ★★★☆☆ For a certain eccentric panache |
| ORIGINALITY: | ★★★☆☆ |
| USEFULNESS: | ☆☆☆☆☆ |
| NERD APPEAL: | ☆☆☆☆☆ |
| ESTIMATED LONGEVITY: | Primarily the domain of gypsies and archly boho types. |

*'The car that runs on oats.'*

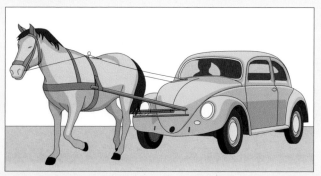

Left: *Laughable as a horse-drawn car seems, it might actually stack up as a form of low carbon emission transport.*

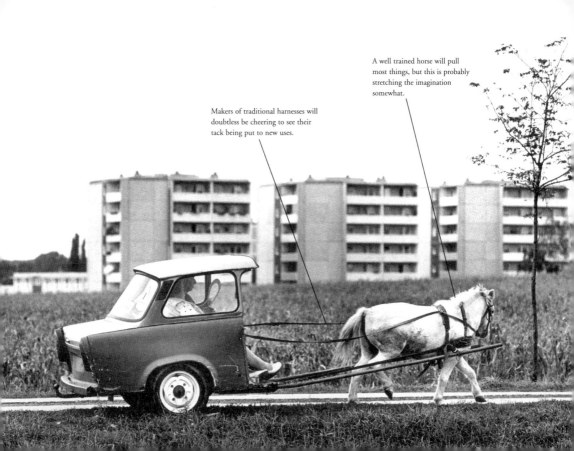

A well trained horse will pull most things, but this is probably stretching the imagination somewhat.

Makers of traditional harnesses will doubtless be cheering to see their tack being put to new uses.

# JET PACK

The jet packer was the larger-than-life, dare-devil hero of many a comic book. Even though using jet power as a means of human propulsion is a highly dangerous activity, it hasn't stopped people from giving it a try. In the 1950s, a couple of American engineers came up with a pack using compressed nitrogen which, it was claimed, would enable you to run at up to

50km/h (30mph). It never got off the drawing board. Further efforts, this time powered by hydrogen peroxide followed, but again had to be abandoned, owing to the assured flying time being barely thirty seconds. This, and the dangers of flying anything below average parachute height, have ensured jet packs remain nothing more than stunts – just like in the comic books, in fact.

**RATINGS**

STYLE COUNT: ★★☆☆☆
ORIGINALITY: ★★★☆☆
USEFULNESS: ☆☆☆☆☆
NERD APPEAL: ★★★★☆
LONGEVITY: Another one where the dream will always outshine the reality.

*'Don't try this at home.'*

Left: *The jet packer has a vivid image in the popular imagination, but despite years of experimentation, remains solely the domain of stuntmen.*

These jet packs are relatively cheap to make. Fair enough, given that the amount of time you can expect to spend airborne is probably under thirty seconds.

The packs can work in outer space where the absence of gravitational pull means less thrust is required.

# *LAMBORGHINI ESPADA*

The very name Lamborghini sizzles with speedy expectation, but the Espada was a misguided attempt by the Italian company to come up with an exotic four-seater. It kept some of the classic speedster attributes – a blazing V12 engine, just two doors and a fastback profile virtually unprecedented for so large a car. But if the Espada was a head turner on the road, internally it was a mess. A complex switchgear arrangement, the earth points prone to corrosion, had you reaching across the passenger seat to get to the headlight switch. The double glass doors were easily shattered, and the low-slung backseats made for an uncomfortable ride for passengers. Lamborghini eventually returned to doing what it does best, which was producing zippy two-seaters like the Miura.

*'When you shifted gears it felt like you were riding a pogo stick.'*

Left: *Espada derived its name from the sword used by Spanish bullfighters, and you couldn't fault its V12 engine, which indeed roared like a tormented bull.*

With the Espada, Lamborghini was attempting to build on the success of the Miura, which had set new standards for high performance two-seaters. But designing a good touring car proved a bridge too far.

There was no arguing with the Espada's road presence, even if the double headlights looked a little ugly.

# METHANE-POWERED CAR

At first glance, Harold Bate was just another batty, wild-haired British eccentric. When he explained how he had been running his old Hillman for years on nothing more than methane gas extracted from chicken and pig manure on his farm, everyone laughed. Yet Harold knew what he was doing, and just a couple of buckets of manure, a tin drum and a carburetor converter enabled him to save vast sums on fuel. He was even able to heat his farm buildings using the method, which was, effectively, a low-carbon, renewable energy.

But Harold's idea had no chance of mass uptake. Not only did it need

constant access to large amounts of manure, it took a good half an hour of pumping away on the extractor machine to squeeze out something equivalent to 30l (7gal) of fuel. For the initial extraction method to be effective, the methane digester had to be kept at a steady temperature of around 26°C (80°F), which was not too easy in an old barn in the chill of an English winter.

**RATINGS:**
STYLE COUNT: ★★★★★
ORIGINALITY: ★★★★★
USEFULNESS: ★★★☆☆
NERD APPEAL: ★★★★★
LONGEVITY: Bate marketed his carburetor converter with some minor success, and his ideas don't seem quite so ridiculous today.

## 'Put a chicken in your tank.'

Left: *Harold Bate's ideas seemed crazy in the 1960s and 1970s, and his approach was certainly eccentric. However, his idea of making fuel from the by-products of animal waste has much in common with current ideas on alternative sources of energy.*

Harold Bate's method was simple. He sealed 18–23 l (4–5 gal) of chicken or pig manure in a drum and heated it to a constant 26°C (80°F) with a small kerosene lamp. But the gas, collected in bottles, took an age to pump out.

It was certainly cheap, clean fuel, and you weren't emitting huge amounts of carbon monoxide.

6844 FH

247

# MONOWHEEL

While no one takes the monowheel seriously as a means of transportation today, early pioneers approached the matter with due sobriety. To properly qualify as a monowheel, the rider must sit inside the wheel, rather than on top. The big outer wheel is usually driven by a smaller wheel within it. The early wheels were sometimes pulled by horses, or relied on pedal power, but by the beginning of the 20th century they were generally powered by engines. Great claims were made for the speed some of them could achieve, but the huge dependency on the balance skills of riders meant they were never more than gimmicks.

Despite this, monowheel enthusiasts still engineer their own variants today, using imaginatively recycled parts from an array of machinery from airplanes to tractors. While record speeds of around 80km/h (50mph) have been attained, however, they are still regarded as no more than a comic spectacle.

**RATINGS**

**STYLE COUNT:** ★★☆☆☆
**ORIGINALITY:** ★★★☆☆
**USEFULNESS:** ☆☆☆☆☆
**NERD APPEAL:** ★★★★☆
**LONGEVITY:** Zero as a serious means of transport, but endless as a mechanical freak's plaything.

*'It's fine, providing you don't have to turn any corners.'*

Left: *While most monowheels today are engine driven, some enthusiasts retain an affection for traditional effects. This one has a small engine, but also possesses the handlebars of a conventional bike.*

Increasing engine power means speeds of more than 80km/h (50mph) have been clocked on the latest monowheels.

While the rider sits inside the wheel, his seat is usually tilted to one side as an aid to balance.

249

# NASH METROPOLITAN

The Metropolitan was a brave bit of counter-intuition in an American car industry of the 1950s wedded to the idea of 'bigger is better'. What the public really wanted, the Nash company reasoned, was a small, inexpensive model, as long as it had traditionally streamlined American bodywork. So, using the small-car expertise of British manufacturers Austin, the Metropolitan combined a tiddly horsepower and an engine barely one-quarter the size of a big Chevrolet, with a sporty profile. It was available in two-tone, and as hardtop or a convertible. The result was something of a flop. While you could learn to love the appearance, the car had an uncomfortably high steering wheel and a huge turning circle, and was a nightmare to handle at anything but very low speeds. In addition, the trunk was only accessible if you pulled down the back seat. Consumers in the USA were not impressed.

*'Dangerous at anything above perambulator speed.'*

IMPORTED
Metropolitan "1500"

"Luxury in Miniature"

Left: *America's first real compact car may have been a commercial failure, but time has been kinder, and two-top convertibles look highly distinctive, with their zig-zag stainless steel side trims and white sidewall tires.*

250

Early models had no opening to the trunk unless you pulled down the rear seat, not helpful in a car being aimed at middle America.

For a car about town, a huge turning circle meant that you were holding your breath when it came to parking situations.

USL 545

# OTTO CYCLE

One of the problems of early bike design was the inherent danger involved in sitting so high above the wheel. The Otto Dicycle, patented by ECT Otto between 1879 and 1881, was a strange attempt at providing a solution. The saddle sat just above the middle of two very large, parallel wheels, while the cranks turned the wheels by pulleys and spring-loaded steel belts. Steering was achieved by slackening the belts so that one wheel went around faster than the other. While going downhill was conceded to be tricky, it was claimed that learning to balance was achievable within minutes. Apart from a faithful number of 'Ottoists', as riders of the bikes were called, the wider public was less convinced.

**RATINGS**

| | |
|---|---|
| **STYLE COUNT:** | ★☆☆☆☆ To look at, not to ride |
| **ORIGINALITY:** | ★★★☆☆ |
| **USEFULNESS:** | ☆☆☆☆☆ |
| **NERD APPEAL:** | ★★★★★ |
| **LONGEVITY:** | Scarcely saw out the 1880s. |

*'Dicycling with death.'*

Left: *Riding a dicycle must have been a terrifying experience. A tiny little 'stalk' on a wheel at the back was provided to help stop you from falling off backward, but the chances of being hurled forward while going downhill were immense.*

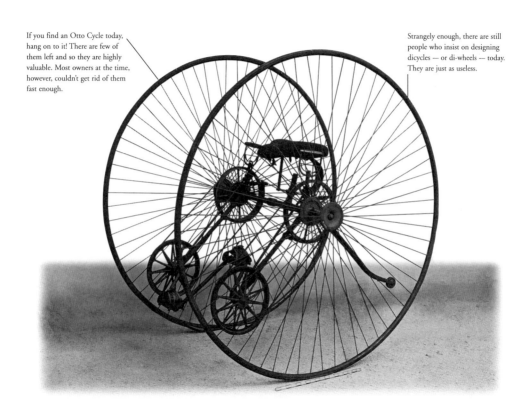

If you find an Otto Cycle today, hang on to it! There are few of them left and so they are highly valuable. Most owners at the time, however, couldn't get rid of them fast enough.

Strangely enough, there are still people who insist on designing dicycles — or di-wheels — today. They are just as useless.

253

# *PEDELUXE CYCLE CAR*

The pedal-powered car, whose heyday ran from the 1920s to the late 1930s, combined the mechanics of a tricycle with the bodywork of a very small car. The advantages of this car were that it provided protection from the elements and accidents and, of course, that it was inexpensive to run. They provided a good way to get some exercise too.

But at a time when motor fuel was still cheap and plentiful, and no one had heard of climate change, the public had fewer concerns about fuel economy. In any case, these light-bodied vehicles handled poorly on the rough-and-ready roads of that era. As large car manufacturers like Ford were

increasingly able to offer cheaply powered motor cars, the pedal-powered car declined in popularity, and those that did survive tended to replace pedal-power with engines.

## *'The mechanics of a tricycle in the body of a car.'*

*Left: Sidney Whitehead's Pedeluxe looked a little lightweight, but if you were nervous about driving an engine-powered car in those still relatively early days, it probably represented a good compromise.*

254

At least you never had to worry about making sure the tank was full before you left on a long journey. However, the weight of pedalling the car probably meant you never used it to go much further than the shops.

An environmentally friendly car from a time when people weren't concerned about cars being environmentally friendly.

# PENNY FARTHING

Penny farthings may look ridiculous, but when they first came out in the 1870s, they represented 'progress'. Their predecessors were made of wood, had metal tires and were so uncomfortable they were nicknamed 'boneshakers'. While the penny farthing still had its pedals attached to the front wheel, the latter was much larger because it was now understood that more ground would be covered with each rotation of the pedals.

But they were inherently hazardous vehicles to ride. Mounting and

dismounting was a precarious business and, with the rider sitting directly above the front wheel, it was easy to be pitched forward over the handlebars if you hit a stone in the road or had to brake fast.

**RATINGS**

| | |
|---|---|
| STYLE COUNT: | ★★★★☆ The must-have item of the 1880s man about town. |
| ORIGINALITY: | ★★★★☆ |
| USEFULNESS: | ★★★☆☆ |
| NERD APPEAL: | ★★★★★ |
| LONGEVITY: | Just a handful of years before being consigned to curiosity value only. |

*'They're a great way of having a peak through people's front windows.'*

Left: *When penny farthings first came out, they were the height of fashion among young men with plenty of money. But they were quickly made obsolete by smaller, more comfortable, bicycles with pneumatic, as opposed to solid rubber, tires.*

Originally called ordinaries, they became known as penny farthings as a form of derision in the 1890s. The term comes from the large English penny and the smaller quarter penny, known as a farthing.

Despite their precariousness, great feats were achieved on them. In 1884, an Englishman, Thomas Stevens, cycled from San Francisco to Boston on one, and then set off to cycle around the world.

# PURVES'S DYNASPHERE

Unlike most monowheels, which tended to be diminutive in size, Dr J.H. Purves's Dynasphere was so huge that, had it proven roadworthy, it would have dwarfed most of the cars of the day. Weighing 450kg (1000lb), its engine and the little compartment in which the driver sat formed a unit revolving on rails within a spherical cage, which remained upright as the wheel was driven forward. The grid-like tire construction gave the driver full vision of the road ahead and there was a little canopy over his head for cover from the elements. Dr Purves's tried it out on the beaches of southwestern England, and even got as far as a run on the famous Brooklands race track in 1932. The Dynasphere proved so cumbersome to steer, however, it nearly ran someone over. It seems Dr Purves lost interest in the idea shortly thereafter, and the Dynasphere was never head of again.

*'A wheely big invention.'*

Left: *You wouldn't have wanted to get in the way of the dynasphere when it was on the go, but one person did at Brooklands race track, and nearly paid with their life.*

258

Dr Purves's wife had a go at driving the wheel, too, but found it no easier to handle.

The plan was for a glassed-in cabin which would seat at least five people, along with headlights and a bumper, to be added, but Dr Purves seemed to lose heart after the Brooklands fright.

259

# RELIANT ROBIN

I f ever there was a car in which to be seen meant certain humiliation, it was the Reliant Robin. The poor man's car *sine qua non*, in spirit it was more akin to a motorcycle with a sidecar, and you only needed a motorbike license to drive it. Along with its three wheels, its bodywork was made of fiberglass, enhancing durability and making it less prone to rust, and it was, unsurprisingly, economic to run. For a few years it became a slightly dubious British icon, insultingly called the Plastic Pig, and the source of easy laughs for comedians.

## RATINGS

**STYLE COUNT:** ★★★★★
**ORIGINALITY:** ★★☆☆☆
**USEFULNESS:** ★★★★☆
**NERD APPEAL:** ★★★☆☆
**LONGEVITY:** Proved a surprisingly durable laughing stock through the 1970s and 1980s.

*'You could get 145km/h (90mph) out of it, but it meant the front wheel rose off the ground.'*

*Left: The Reliant was soul-destroyingly ugly, but highly practical, with room for Dad, Mom, three small kids and a dog.*

It was a mistake to compare it with the Mini, a small car that could take corners at speed, while the Reliant was more likely to topple over.

The fiberglass body was a considerable plus, requiring little maintenance and this, added to its light weight and low fuel consumption, made the cars cheap to run.

*Robin* 65

# RENAULT AVANTIME

The Renault Avantime looked like a bulky big van, but had a fatally cramped interior. It was designed by Matra, a coach-building affiliate of Renault, and contained huge amounts of coach-like tinted glass. This included two sun roofs, and electric side windows which, when lowered, opened the entire side of the car, rendering a countryside cruise a gusty affair. Worse was that, while it was supposed to seat four people, it only had two huge, double-hinged doors, making it a nightmare to get in and out of in tight parking situations. When Matra pulled out of automobile construction in 2003, Renault had no hesitation in halting production of what was proving to be a disastrously poor seller.

*'Unconventional, unprecedented, uncomfortable – and unsold.'*

**Left:** *Although the Avantime looked long, footspace was cramped. Passengers in the back had no room to stretch, and in the front seat the passenger had to find space alongside the air conditioner, which protruded into the footwell.*

The car's upright driving position took some getting used to, and made the vehicle feel heavy and cumbersome.

Not exactly stylish, its sheer size undoubtedly gave it road presence.

263

# ROCKET-POWERED BICYCLE

In 1931, a German named Herr Richter, perhaps having spent his summer holidays reading too many boys' comics, became fixated with the idea of setting a new world speed on a rocket-powered bike. All this seems to have involved was donning a long raincoat and a serious-looking racing driver's helmet and goggles, and hopping onto the back of an ordinary pedal bike with 12 solid-fuel rockets strapped to the rack. A white box held a battery, which provided the ignition, and was controlled by Herr Richter via switches on the handlebar.

Everything seemed to be going well as the bike puttered along the track, but when the good Herr pushed the speed up to 80km/h (50mph), the engine began to overheat. A mighty explosion suddenly sent up a cloud of dust from which, moments later, a dazed Herr Richter was seen emerging, helmet askew, clambering out of a ditch. He seemed unhurt but, to his great disappointment, the bicycle was a hissing and steaming wreck.

*'A Herr-raising experience.'*

**Left:** *It was not unusual to experiment with rocket power in the 1920s and 1930s, but trying it out on a bicycle was going a bit far.*

Unfortunately, his assistant rather let the side down by turning up in a scruffy, all-purpose mechanic's overall.

The good Herr certainly took it all very seriously, donning a very natty submarine-style commander's raincoat. You have to hope it didn't get ruined when he landed in the ditch.

# SCHEUER'S TRANSPARENT ROWING BOAT

As new materials like fiberglass began to be used in boat building starting in the middle of the twentieth century, the idea of transparent, or so-called 'glass-bottomed' designs became a possibility. They afforded rowers or passengers wonderful new windows into the world of aquatic wildlife and waterscapes hitherto available only to divers.

Most of these early designs, however, proved unsuitable. The structural members were made of wood or aluminum, with plastic used to form the transparent skin of the boat. But cracks would quickly appear between the materials and sprout leaks, and the plastic would soon cloud up, compromising the 'transparent' view. Scheuer's rowing boat would doubtless have suffered from all of these problems and, in any case, given its evident fragility, it's unlikely it would have provided much insight into sea life beyond that found in the average garden pond.

*'Clouded visions.'*

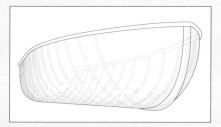

Left: *A plastic rowing boat you could see through wasn't much use on a muddy pond either.*

Scheuer's boat had neither the sturdiness of a normal rowing boat, nor the light handling qualities of a canoe. It looks as if it could have capsized at any moment.

Soon the plastic would have clouded up and the 'view' would have been as interesting as looking into a filthy watertank.

# SHAPSTON AQUAPLANE

A quaplaning to most people in the 1930s meant some game individual standing on a surfboard and skimming across shallow waters while being dragged along on the end of a car or boat tow rope. That evidently wasn't good enough for the inventor of the Shapston Aquaplane, seemingly tested on the beach at Santa Monica, California, in 1935.

The intention was for the user to lie on the aquaplane board, creating forward motion by turning the cranks that revolved a small propeller at the stern end. A distinctly optimistic speed of 12 knots was forecast, and it was hoped that local lifeguards would be incorporating it into their collection of rescue gear before the year was out. But it's hard to imagine the trial of this most unlikely looking of gadgets achieving anything other than farcical results.

*'It'll certainly save a life if you don't use it.'*

Left: *Aquaplaning was a 'fun only' activity in the 1930s. But while the inventor of the Shapston Aquaplane had the commendable intention of adapting it to save lives, his idea proved to be all at sea.*

It was a nice sunny day when the experiment took place, but the mood among the participants would soon become overcast as the undeniable hopelessness of the device became apparent.

Buoyancy was to be achieved by these two large tanks, made of bronze and supporting a weight of up to 80kg (180lb).

# SINCLAIR C5

Written off by the press as a joke when launched in 1985, even today mention of the Sinclair C5 still provokes derisive laughter. It was intended to start a revolution on the roads as the first in a fleet of battery-powered electric cars, but it quickly ended in tears, with only around 1200 models sold, and its inventor Sir Clive Sinclair's company heading into receivership. The C5 was really nothing more than a tricycle, and you didn't even need a driver's license to operate one because it couldn't go faster than 24km/h (15mph). Its open design was hopelessly unsuitable for the wet, cold, windy British climate and, being so low to the ground, serious doubts were raised about its safety on the country's juggernaut-clogged highways. Not even getting formula one racing legend Stirling Moss on board for the publicity campaign could save this wretched little vehicle.

## RATINGS

**STYLE COUNT:** ★★★★★
**ORIGINALITY:** ★★★☆☆
**USEFULNESS:** ★☆☆☆☆
**NERD APPEAL:** ★★★☆☆
**LONGEVITY:** 10 months from launch to bankruptcy.

*'Handy for shifting boxes around in warehouses, but rubbish anywhere else.'*

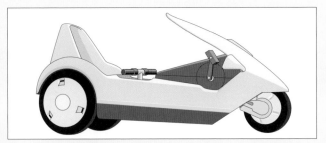

Left: *It took just one look at the C5 to see how vulnerable it would be on the road.*

A modification was hastily introduced, equipping it with a safety reflector screen, but it was too late to undo the damage.

It was launched in a blaze of publicity, but the press tore it to shreds, one cartoon portraying it driven by lemmings hurtling into a collision with a juggernaut.

# SOLAR-POWERED BICYCLE

A solar-powered bicycle, if it ever worked, would require less leg work on the part of the rider, and could even enable them to travel faster. So far, however, despite many attempts, no one has come up with one that works in practice as well as in theory. Most efforts involve placing a solar panel on the front or back of the bike which, using the suns rays, powers a battery mechanism. But all experiments so far have failed to tackle the same problem, namely, it would take hours of sunshine to derive enough energy to power the battery for even a short ride. In any case, the bicycle is already one of the greenest forms of transport available, so why try to make it any more so?

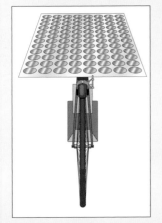

## RATINGS

**STYLE COUNT:** ★★★★★
**ORIGINALITY:** ★★★☆☆
**USEFULNESS:** ★★★★★
**NERD APPEAL:** ★★★☆☆
**LONGEVITY:** So far no one has managed to make this one work, but it may be simply a matter of time.

*'It only works when the sun's out.'*

Left: *The enormous size of the solar panel is not looking good. These bikes simply don't work without inordinate amounts of sunshine to power the battery.*

The rider has to contend with a battery stuck between his legs.

By positioning the rider lower, space is made for a larger solar panel.

273

# SUITCASE CAR

One of the real drags of arriving at an airport is waiting for a taxi, or finding some other means of transport. How convenient would it be to open up your suitcase, take out your car, and simply drive off? Such was the ideal world the designers of the suitcase car sought to create. Powered by a 40cc (2.4cu in) two-stroke engine and holding enough fuel for two hours

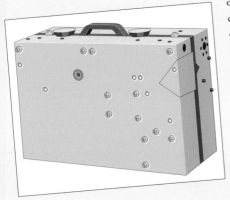

of driving, it could cruise along at a stately 43km/h (27mph). The 'car', which took seconds to assemble, had a chassis, brake-lights and even headlights for nighttime driving. However, the original was destroyed in a parking accident shortly after it was designed in the early 1990s, probably by a driver who wasn't even aware they'd run over it. In these days of stringent airport regulations, of course, you'd never get such a contraption through security.

### 'Don't forget to pack the car.'

Left: *Packing your car in your suitcase seems convenient, but it must have been heavy to carry around.*

Handlebars were used for steering, instead of a steering wheel.

Sales representatives turning up for a meeting on one of these will not have done much to enhance their company's status.

275

# TANDEM BICYCLE

Tandem bikes have always had a faintly geeky air. While long associated with a sentimental old song about a man and his girl Daisy on a 'bicycle made for two', modern technology has turned them into high-performance affairs, capable of reaching great speeds. The trouble is they come with a lot of terminological baggage and riding etiquette that makes them an argument waiting to happen. Not for nothing have they been termed 'divorce machines'.

The great thing about an ordinary pedal bike is its get-up-and-go quality. But with tandems, the person on the front must act as 'captain' and do the steering and who will invariably end up doing most of the pedalling. Going uphill, the two riders must cooperate and coordinate their pedal power. The sheer physical strength needed to act as captain means it's usually the male

who rides up front, barking orders while the female, acting as 'stoker', seethes behind him with nothing but his big fat posterior ahead of her for hours on end. It is a surefire setting for marital disaster.

*'Say hello to a tandem, and goodbye to your relationship.'*

Left: *In terms of performance, modern tandem bikes are a match for ordinary bikes. But all the technical improvements in the world can't make up for the fact that they can stretch relationships to breaking point.*

In the right situation, especially when riding downhill or on flat ground, two well coordinated tandem riders can actually go faster than a solo bike.

For most couples, it's a question of trundling along with the man in front. Sometimes the fact that he's 'captain' will go to his head.

277

# *TRABANT*

For a car with such dreadful performance levels, the Trabant did remarkably well to endure for more than 30 years, but then its homeland *was* East Germany. The Trabant became a symbol of all that the West liked to think was bad about Communism. If you ordered a Trabant from the factory, it took 15 years to arrive, and when it did what you got was an incredibly noisy, smoke-belching vehicle. Because the East Germans refused to import steel, it wasn't even made from metal, but from a plastic fiberglass called Duraplast. When the Berlin Wall fell in 1989, production continued for a couple more years, but with access to bigger, more efficient Western car models, its days were numbered. Sure enough, production ceased in 1991.

**RATINGS**

**STYLE COUNT:** ★★★☆☆ Loveable ugly ducklings.

**ORIGINALITY:** ★★★☆☆

**USEFULNESS:** ★★★★☆

**NERD APPEAL:** ★★★★★

**LONGEVITY:** The first in the series was produced in 1957, the last appeared in 1991. Difficult to argue with that.

## *'The return of the steam engine.'*

Left: *For all its many shortcomings, the Trabant was cheap to run and had such simple mechanisms there was little that could go wrong. This was just as well, given the delivery of cars and spares in East Germany was measurable in years, not months.*

Given their longevity and symbolism of an era, Trabant car clubs have been set up all over the world since production ceased in 1991.

Trabants had two-stroke engines, and filling the tank, which only had a 27-l (6-gal) capacity, was laborious. You had to lift the hood and, after you'd added the fuel, add two-stroke oil.

# TRANS-ETHER STEAM FLYING MACHINE

In the mid-19th century, as scientists devoted ever more time to solving man's quest to fly, they began to better understand the 'heavier-than-air' principle (for a flying object to move forward, it had to be more powerful than the air flows dragging or pushing against it). This set a rash of would-be

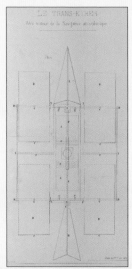

inventors to work, and a flurry of madcap schemes were unveiled. One was the idea of an obscure Frenchman, F. Ducroz, for a Trans-Ether steam flying machine.

It looks as incredible as its name suggests, but like many of these other steam-powered contraptions, it could never have left the ground. Still, you have to give him full marks for imagination. With four monstrous, steam-driven paddlewheels, it looks like something out of a Jules Verne novel. Had the movies been around in those days, Monsieur Ducroz might have landed himself a contract as a Hollywood designer. But he would have had to take a boat to get there.

*'It was to be a ship of the sky, but would have been more use as a steamboat on the Mississippi.'*

Left: *In the 1850s, the idea of aerial steam carriages was not uncommon. However, these early designs were just too heavy to provide effective lift in relation to the great weight of the machine.*

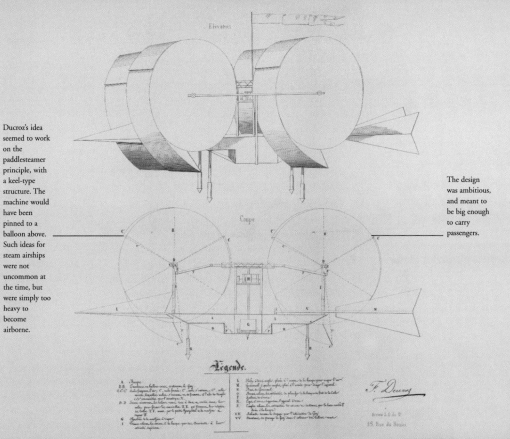

Ducroz's idea seemed to work on the paddlesteamer principle, with a keel-type structure. The machine would have been pinned to a balloon above. Such ideas for steam airships were not uncommon at the time, but were simply too heavy to become airborne.

The design was ambitious, and meant to be big enough to carry passengers.

Elévation

Coupe

Légende.

F. Ducroz

Bureau S.G. du St

35, Rue du Bassin.

# UNICYCLE MOTORBIKE

The Uno isn't really a unicycle at all because it has two wheels, though they are paired together to create the illusion of being one. Still, the experience of riding it would be just as scary as riding a unicycle. The concept behind it was to create a non-polluting electric alternative to the motorbike. It's certainly economical to run. It is so light and small it can be taken indoors to be charged up off the mains. But this odd vehicle has only a single on/off switch for controls. The rest is entirely dependent upon the rider's powers of balance. This means that, while it can allegedly 'turn on a dime', only the most intrepid will dare to mount one.

**RATINGS**

| | |
|---|---|
| **STYLE COUNT:** | ★★★★★ |
| **ORIGINALITY:** | ★★★★★ |
| **USEFULNESS:** | ★★★★★ |
| **NERD APPEAL:** | ★★★★★ |
| **LONGEVITY:** | When first announced, it ignited some interest, but it's difficult to see any practical application for it. |

*'The Uno that's a duo.'*

Left: *The Uno actually has two wheels, but still places immense demands on the rider's sense of balance. There are no brakes, making sudden stops problematic.*

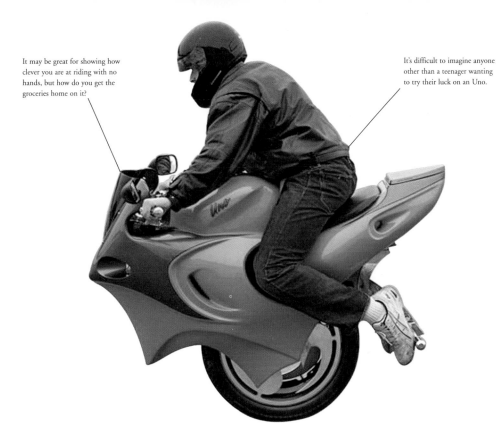

It may be great for showing how clever you are at riding with no hands, but how do you get the groceries home on it?

It's difficult to imagine anyone other than a teenager wanting to try their luck on an Uno.

283

# VOLK'S DADDY LONG LEGS

Magnus Volk's narrow gauge electric railroad in the English seaside town of Brighton first opened in 1883. It still survives as one of the oldest electric railroads in the world. But Volk's subsequent plan to build an extension was a bizarre affair that ended in failure. Running the railroad directly beside a chalk cliff would have left passengers vulnerable to rocks falling from above, so he took the rash step of running the track through the sea, meaning that it was completed immersed at high tides.

The conveyance for the passengers was an odd-looking 45.7-tonne (45-ton) tram on tubular legs 7m (23ft) high. It quickly became known as Daddy Long Legs, and at first seemed quite popular. The trouble was that, even at low tides, water often ran over the line, slowing the tram's progress,

and at high tides it seized up completely. Frequent safety works to the sea walls, which put it of action at key points in the holiday seasons, further undermined profits, and within three years of its opening the line was closed. The Daddy Longs Legs stood forlorn for a time, tethered to a pier, before its rusting remains were towed away for scrap.

*'A sea voyage on wheels.'*

Left: 'Daddy Long Legs' was described as a cross between an open top streetcar, a pleasure yacht and a seaside pier, but its success was hampered by the impracticality of running a railroad so close to the sea.

People were enthusiastic when it first began taking passengers, but they grew impatient when high tides slowed it to a standstill.

The Lifeboat dinghies were an ominous sign. If the tram was caught in a sudden storm whipping up the sea, there was no means of getting back to dry land.

285

# WATER BIKE

In most countries, if you were seen trying to float a bicycle across water you would probably be certified a lunatic, but not, it seems, in China where any madcap do-it-yourself activity is likely to be keenly evaluated by financially stretched citizens. The water bike is equipped with 8 water pontoons. These are raised up when you reach dry land, enabling the bike to revert to conventional mode. Vanes attached to the rear spokes reduce the swirl of the water flows. It doesn't seem to make much sense, but if you live in a tiny little village cut off by water from the glittering lights of the city, maybe anything's worth a try.

**RATINGS**

**STYLE COUNT:** ☆☆☆☆☆
**ORIGINALITY:** ★★★★☆
**USEFULNESS:** ☆☆☆☆☆
**NERD APPEAL:** ★☆☆☆☆
**LONGEVITY:** For Chinese do-it-yourself freaks only.

*'Cycling around on the river.'*

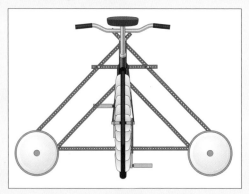

Left: *The prototype diagram of the water bike looks simple enough, but would you fancy riding it across a river?*

You hate to sneer at attempts to fashion something out of nothing, but wouldn't it have been less bother to invest in even the most modest of rowing boats?

The bike looks seaworthy with its pontoons, but how would it cope with the backwash from larger boats?

287

# WONDERBIKE

There's an unspoken supposition behind inventions that they must have a real purpose or some practical end use. Even if the outcome has very little chance of success, this very purposefulness commands a certain respect, however tenuously merited.

But why does this have to be so? What's wrong with an invention that had no purpose whatsoever right from the very beginning? Mr Glubule's Wonderbike, for instance, is obviously meaningless. It's highly likely that not a single intelligent thought passed through his head as he was happily assembling its miscellany of unconnected, irrelevant objects. It's just a bike, with a couple of inflatables clamped on the side, a pathetic little kiddie's windmill hanging off the end, and a strange sort of trampoline thing that presumably serves as a wonky kind of rack or bed. But it's a statement of some kind for Mr Glubule, and he's clearly derived a lot of pleasure making it, causing no one any harm. Who are we to sneer?

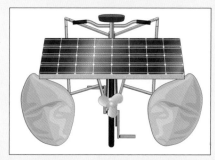

*'I call it my wonderbike because people will wonder just exactly what it's for.'*

Left: *First impressions may suggest that the wonderbike is ingenious, but in fact it's just impractical.*

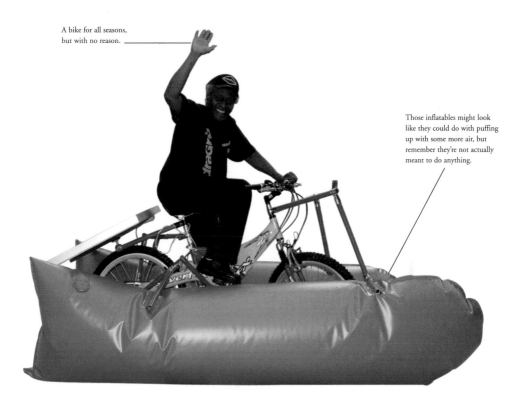

A bike for all seasons, but with no reason.

Those inflatables might look like they could do with puffing up with some more air, but remember they're not actually meant to do anything.

289

# GIZMOS AND GADGETS

Unlike the dubious inventions of earlier chapters, this section does feature a few genuine attempts at solving problems in order to make life easier. You may hold your nose about pay toilets and portable female urinals, but they have undoubtedly been useful in answering nature's call. The post office air torpedo probably scattered more letters to the wind than it delivered, but it was a seriously attempt that was undertaken to speed communications up. A pocket chainsaw seems absolutely insane when marketing is aimed at the hiker, but it has definite claims as an outdoor accessory. Still, there's plenty of nonsense here, too. The creator of the hunting skis was one sandwich short of a picnic, clearly. So, too, was the inventor of the parachute overcoat. Unfortunately, his pipe dream came to a truly horrible end before a horrified crowd of onlookers below the Eiffel Tower in Paris.

Left: *Torpedo post took the idea of airmail into new territory, though seldom quite where the sender intended it to go.*

# DIGITAL ROSARY

Visit the gift shop of any religious site with big tourist appeal and the worthless, exploitative material on sale will boggle your mind with its sheer tackiness. So no one should be surprised at the arrival of the digital rosary. At the press of a button on this 'precious portable item', a female voice recites the Hail Mary, with an answering female choir in the background. It's a horrible idea – a convenience prayer machine – that non-believers and cynics will simply say frees up its users to carry on with the daily business of sinning.

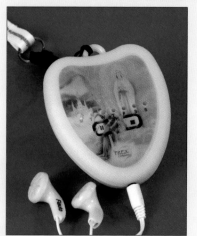

**'Rosary beads that look more like egg-timers.'**

Left: *An intelligent control button works recordings that recite the four mysteries of the rosary. A female voice and choir softly intone the prayers. Handy for the disabled perhaps, but for anyone else, simply a scam.*

Saying the rosary while touching the beads was supposed to be about contemplating the mysteries of faith, hardly something reducible to pre-recorded messages.

The rosary doesn't look very spiritual, but then this has nothing to do with prayer, and everything to do with commerce.

# ELECTRIC SHOPPING CART

Supermarkets have done their utmost to eliminate shopper exertion, but they've never quite been able to do away with the tiresome matter of having to walk up and down the aisles to collect your groceries. But whoever came up with the idea of the electric shopping cart should be hauled away in chains. Visiting the store is hazardous enough for the lone shopper equipped only with a small hand-held basket, negotiating their way between aisle-hogging monsters wielding four-wheeled shopping carts. Letting such bullies loose with coin-operated motorized ones is likely to turn a simple trip to the supermarket into a massacre.

**RATINGS**

STYLE COUNT: ★★★★★
ORIGINALITY: ★★★★★
USEFULNESS: ★★★★★
NERD APPEAL: ★★★★★
LONGEVITY: None. It's just too awful to contemplate.

*'Dodgem cars at the supermarket.'*

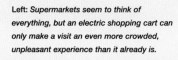

Left: *Supermarkets seem to think of everything, but an electric shopping cart can only make a visit an even more crowded, unpleasant experience than it already is.*

Put some people behind a shopping cart and they think they are king of the road. Can you imagine what they would be like on a motorized one?

There is even talk of adding gears, after which it will only be a matter of time before some fool attempts to ride one over the deli counter.

# EXERCISE CHAIR AT WORK

At first, the office exercise chair looks like a perfectly harmless, comfortable chair. But at the press of a button, it can be converted into a multipurpose gym apparatus. The back cover pulls off to reveal the full horror of a mighty fixed-weight machine, supplied with weights ranging up to 50kg (110lb).

Can you imagine what an instrument of torture this could be if you have a tyrannical boss who suddenly takes it into their head that the staff need to get in shape? There you are, sitting at your desk when in the mad sadist walks, hauls up the chair's tubular frame to its full height, and starts barking out instructions for lateral pulls, bench presses and leg extensions. It's a terrifying thought.

*'If you thought life at the office was tough, it's about to get tougher.'*

Left: *The exercise chair, by encouraging office workouts, is supposed to help reduce stress. But if it gets into the hands of a dictatorial boss, it could actually cause more tension among the workers.*

This man is clearly his own boss, and can exercise at his own will and, in all probability, in moderation. Nonetheless, in the wrong hands the chair could be dangerous.

Even with him, the signs are there that the chair is having a damaging effect. It must never be permitted to get out into the public domain.

# *FRESH KISS BREATH DETECTOR*

Nobody wants to hear they have bad breath, and few people are likely to tell them. A breath detector reminding you of the fact may not be very welcome, but you have to admit, it could come in handy. These detectors will rate your breath up to five different levels, from 'kissable' to a rather scary 'deadly'. They are also small enough for a discreet check to be done without it looking anything more suspicious than a chat on a phone. It could still be something of a passion killer, though, so taking a reading well in advance is advisable.

**RATINGS**

STYLE COUNT: ★★★★★
ORIGINALITY: ★★★★★ (an innovative breath of fresh air)
USEFULNESS: ★★★★★
NERD APPEAL: ★★★★★
LONGEVITY: Definite possibilities.

*'Kissing with confidence.'*

**Left:** *Your best friend won't tell, the saying goes....but a breath detector might.*

Making it look like a phone was a smart move.

If the monitor gives you the maximum reading of 5, you probably already have so few friends it won't matter anyway.

299

# HUNTING SKIS

Paddle power has been put to good use by mankind since early prehistoric times, but there are always characters who have to go that one step too far. The inventor of the aquatic hunting ski was obviously fond of shooting ducks. But whereas most people would settle for doing it in a boat, he thought he'd try wading around on floating skis. Using the paddles to propel himself forward, he also remembered to provide himself with a little seat, anticipating long hours patiently awaiting his prey. What a hopeless optimist. You can be sure that, before he'd splashed around for more than half a dozen steps, every waterfowl in the district would have been aware of his arrival, and made for sanctuary.

**RATINGS**

STYLE COUNT: ★★★★★
ORIGINALITY: ★★★★★
USEFULNESS: ★★★★★
NERD APPEAL: ★★★★★
LONGEVITY: Strictly for eccentrics.

*'The Adventures of Splash Harry.'*

Left: *Another man came up with his own 'aquatic velocipede', comprising three floats attached to iron rods arching up to a saddle. Again, prospects for a good shoot: zilch.*

There's an old saying that if you can't be a successful hunter, you can at least look like one. Here is a case in point. He had the right kit and a decent rifle, but his chances of success were still zero.

You probably had more chance of drowning than returning home with duck for dinner.

# MIDGEATER

It's been said that the tiny midge can single-handedly put people off visiting what would otherwise be regarded as perfectly civilized parts of the world. There are even folks who've upped sticks and moved house because of 'the problem with the midges'. So when you see someone revving up the midgeater, you know they mean business. This is a serious machine that lures the midges or mosquitoes by wafting out waves of $CO_2$, thus reminding

them of the scent of human breath, and a potential mealtime. A bag mechanism sucks them up, which not only kills them off, but interrupts the breeding cycle. It's all hopelessly short-term, of course, and the insects are soon back in bigger numbers than ever before.

**RATINGS**

STYLE COUNT: ★★★★★
ORIGINALITY: ★★★★★
USEFULNESS: ★★★★★
NERD APPEAL: ★★★★★
LONGEVITY: Anyone stupid enough to think you can exterminate insect populations simply by blowing $CO_2$ in the air will be dumb enough to keep on buying these.

*'The Midge who came for dinner.'*

*Left: Insect bites can be horrible, but if the local climate supports populations as dense as 50 million of them per square hectare, perhaps the spot isn't suitable for humans.*

302

Don't mess with anyone powering up a midgeater. You'll know from the determined look on their face that they are engaged on a Very Important Matter and Not to be Disturbed.

They're probably wasting their time, however, because there's no winning with midges and mosquitoes.

# PARACHUTE OVERCOAT

With awful irony, the press called Franz Reichelt 'the Flying Tailor'. He'd combined an interest in tailoring with early attempts at flight, and come to Paris in 1912 with a garment that was a combination of overcoat and parachute, designed to enable him to float, or fly, through the air. Though he promised the authorities he would do a low-level dummy test first, somehow they permitted him to trial it from the first deck of the Eiffel Tower, 18m (60ft) off the ground.

The all-too-graphic final moments of his life were captured on black and white newsreel. There he stands, a bat-like figure, perched on a tiny chair, clearly having second thoughts. He flaps his arms a few times, and a man in a long coat beside him seems to be cajoling him forward. Then, finally, he slides over the parapet, plummeting like a stone, before a gasping crowd, to his dreadful and instantaneous death.

*'The dummy test of all time.'*

*Left: Reichelt willingly posed for cameras before making his fatal Eiffel Tower parachute overcoat leap, which was also caught on live film.*

Reichelt's death might have been marginally less disturbing if the film had not so clearly shown him having to summon up the courage to leap.

Reichelt's 'parachute' also doubled as a rather bulky overcoat.

# PAY TOILET

We must thank the Roman emperor Vespasian for the first pay toilets, introduced to Rome in the first century AD. Typically, they were leisured affairs, with hot and cold running water and good sewage systems. Europeans cities of the 18th century, however, opted for something less flashy. Purveyors of public toilets wandered the streets. In exchange for some cash, they handed you a bucket and wrapped a large cloak around you while you did the necessaries.

This doesn't make it any easier to love the modern automated pay toilet. The ones that attempt to ration toilet paper, restricting the number of times you can push the button that releases it, are bad enough. But what of the ones where the doors pop open after a set number of minutes to activate the cleaning mechanism? It turns what should be an untroubled private process that you have, after all, *paid for*, into a panic-inducing race against time.

## 'To pee or not to pee?'

Left: *Pay toilets were partly installed to deter drug addicts, the homeless and other social groups who might abuse them. But they have been unable to lose their squalid reputation.*

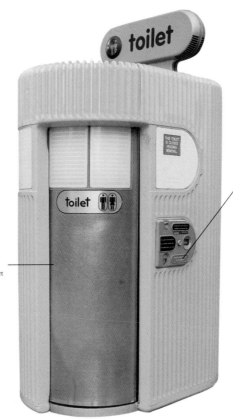

Not the least of the frustrations of automated pay toilets is that they always seem to be 'Out of Order', or unable to provide change.

Because the cleaning and door-opening mechanisms are already set to go off at regular intervals, the toilet paradoxically provides relief and anxiety at the same time.

# PERPETUAL MOTION MACHINE

Perpetual motion machines, which generate energy for free, have fascinated inventors for centuries. Frenchman Villard de Honnecort was making diagrams of perpetual motion wheels in the 13th century, but there is no evidence that he actually got around to putting the idea into practice.

However, it was Robert Fludd, a 17th-century English physicist and mystic, who is widely credited with coming up with the first recorded attempt to describe such a device that could be put to useful work. His 'water screw' perpetual motion machine was illustrated in a wood engraving in 1660. The idea was that water lifted by a pump would drive an overbalanced waterwheel, which in itself would drive a millstone. The device

would then pump the water back into its own supply tank, hence facilitating perpetual motion.

Some have argued that Fludd's drawing was actually an attempt to show the impossibility of such a contraption, but it certainly generated a motion of its own. As late as the 1870s, inventors were still attempting to patent variations on the contraption he described.

*'The Gods of Physics duly smite them/To be reborn ad infinitum.'*

Left: *The laws of thermodynamics indicate that perpetual motion machines are impossible. Yet modern attempts at using solar power or harnessing the forces of wind or water to generate energy are simply more sophisticated developments of the supposedly crazy ideas of our ancestors.*

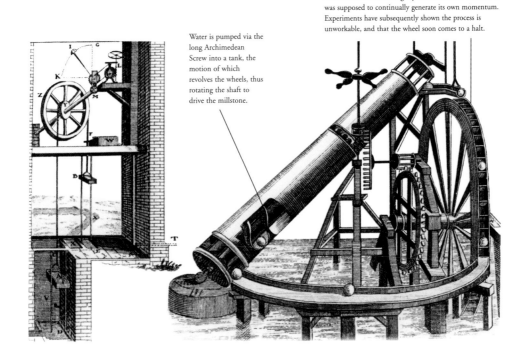

Water is pumped via the long Archimedean Screw into a tank, the motion of which revolves the wheels, thus rotating the shaft to drive the millstone.

Because the wheel was slightly overbalanced to one side, it was supposed to continually generate its own momentum. Experiments have subsequently shown the process is unworkable, and that the wheel soon comes to a halt.

# PLANT SPEAKER SYSTEM

It's unfair to say the Japanese have no imagination, for they certainly can come up with some fanciful ideas. 'Ka-on', meaning 'flower sound', is a gadget that turns plants into audio speakers, allegedly, making the leaves tremble as if in some weird, hypnotic dance. The system operates via a donut-shaped magnet and coil, which you place in the bottom of a plant pot and hook up to your CD player or computer. With the flowers placed in the pot and the stems fed through the hole in the magnet, sound is said to travel up the plant, making the leaves and flowers shake. Quite how this is supposed to be therapeutic is questionable. It certainly can't be much fun for the plants to be jolted around in this way, like unwilling guests at a thrash metal gig.

*'The magnet is more use as a paperweight.'*

Left: *Plants have long been appreciated for their tranquility-inducing, therapeutic qualities. Why not be kind to them in return and just let them grow in peace?*

The manufacturer of the devices says the plants like listening to the music, but since plants can't communicate, how does he know?

The Ka-on is said to send the sound of the music in all directions, rather than the normal single-spot speaker system. But it must be a feeble affair when you can only hear it by placing your ear up close to the leaf.

311

# PORTABLE FEMALE URINAL

Music festivals have enjoyed a surge in popularity in recent years, but a characteristic most share is that, when it comes to restroom provisions, women come second. A new trend, however, is the provision of sectioned-off female-only urinals. Here females are handed a plastic funnel that enables them to urinate while standing up. It alleviates them of the indignity of the need to find a bush to squat behind, and or having to join the lengthy lineups for the male toilets.

The funnels are, of course, reusable, and are augmented by 'go-bags', which are as revolting as they sound. So while they seem like a good idea, it's easy to imagine the already pungent air of a hot summer festival being fortified by a faint whiff of urine wafting from a thousand shoulderbags.

**RATINGS**

| | |
|---|---|
| STYLE COUNT: | ★★★★★ |
| ORIGINALITY: | ★★★★★ |
| USEFULNESS: | ★★★★★ |
| NERD APPEAL: | ★★★★★ |
| LONGEVITY: | When you gotta go, you gotta go, so it's easy to see this meeting popular demand. |

*'Pee, shake and stick it back in your pocket.'*

**Left:** *Restroom provision is usually so poor at outdoor events that women end up adding to the long lineup for the men's.*

Of course, the mini '*pissoires*' are reusable, so it can't be very pleasant having to carry it around in your pocket or bag.

Female portable urinals look pretty undignified, but they're a definite improvement on having to squat behind a bush.

313

# ROCKET MAIL

You can't accuse postal operators of not having tried to speed up services over the years. In the 1930s, attempts were even made to deliver mail by rocket. The first properly recorded effort came from Austria in 1931, when Friedrich Schmiedl launched a rocket containing more than 100 letters. In India, Sidney Smith was doing the same, becoming the first person to send

rocket mail across a river and to launch one carrying a parcel. In Germany, meanwhile, Gerhard Zucker was experimenting with small missiles known as powder-rockets.

In 1934, Zucker staged a demonstration for the Royal Mail, with a canister about 1m (3ft) long and 18cm (5in) in diameter, packed with around 1200 letters. Unfortunately, the rocket exploded in mid air, scattering burning letters all across the ground. The idea of letters delivered by rocket has seldom been taken seriously ever since, although the Russian military is trying out experimental rocket mail launches from submarines.

## RATINGS

STYLE COUNT: ☆☆☆☆☆
ORIGINALITY: ★★★★★
USEFULNESS: ★★☆☆☆
NERD APPEAL: ★★★★☆
LONGEVITY: Improving technology could eventually provide the means for this one.

## 'Flare mail.'

Left: *Rocket mail missiles or canisters were supposed to land by deploying internal parachute systems on arrival, but few ever reached their destination.*

Rocket mail wasn't a total lost cause. One of Zucker's efforts was witnessed to have risen between 400–800m (1300–2625ft) in the air.

When he moved to England, he started worked on another prototype which used, among things, butter as a lubricant.

315

# TORPEDO POST

Germany in the early decades of the 20th century prided itself on being at the cutting of technological innovation, and you have to admit that Berlin engineer Richard Pfautz's plans to dispatch mail across country by torpedo post was certainly blue skies thinking. The mail carrier was a metal "missile", resembling nothing less than a torpedo, driven by air propellers at the front and rear, with motors taking their power from an electrified cable along which the projectile ran. The trouble was that, while it was supposed to be quicker than air mail, and only a little more expensive than conventional post, it required a network of cables if it was to succeed in making deliveries "to the farthermost frontiers of the country."

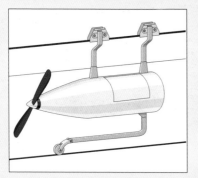

Unsurprisingly, in a country experiencing a severe economic depression in the early 1930s, it never progressed beyond the testing stage.

**RATINGS**

STYLE COUNT: ★★★☆☆
ORIGINALITY: ★★★☆☆
USEFULNESS: ☆☆☆☆☆
NERD APPEAL: ★★★★☆
LONGEVITY: Didn't endure to see the arrival of Hitler's dictatorship.

*'It made airmail look like snail mail.'*

Left: The plan was for torpedo mail carriers to "whizz" the mail along specially constructed cable runways at 200 mph.

In cold weather, frozen points would surely have surely brought the system to a grinding halt.

Berlin engineer Richard Pfautz had high hopes for his new system. But requiring network of cables to be laid out across the country, it wouldn't have come cheap.

# *INDEX*

# Picture Credits